D0188356

PART-TIME FARMING

PART-TIME FARMING

Katie Thear

WARD LOCK LIMITED · LONDON

First published in Great Britain in 1982
by Ward Lock Limited, 82 Gower Street,
London WC1E 6EQ, a Pentos Company.

Layout by Bob Swan
House editor Helen Douglas-Cooper
Text filmset in 11/12 Ehrhardt
by Butler & Tanner Ltd

Printed and bound in Great Britain by
Butler & Tanner Ltd, Frome and London

British Library Cataloguing in Publication Data

Thear, Katie
Part-time farming.
1. Part-time farming
I. Title
338.1 HD1476.A3

ISBN 0-7063-5932-1

Frontispiece: Aylesbury ducks: the traditional table breed in Britain.

Acknowledgments

The author and publishers would like to thank the following for supplying pictures for the book: Batricar/John R. Simmons page 21; British Turkey Federation page 158; Mrs H. Brace page 83 top; *Farmers Weekly* pages 31, 36, 37 top and below, 39, 40, 41, 44, 79, 100, 102 top and below, 117; Alfred Cox (Surgical) Ltd/John Topham Picture Library page 115; Henry Grant page 133; Geoff Goode pages 84, 85; Marc Henrie pages 18, 19, 20 top, 28, 52 top and below, 89, 95 top and below, 99, 104, 119 left, 120, 127, 142, 147, 163, 165, 166; Kubota Tractors page 20 below; Jane Miller frontispiece, pages 8, 27, 71 top and below, 77, 78, 108, 116, 119 right, 136, 145, 150, 153, 155, 184; *Poultry World* pages 128, 130; *Practical Self Sufficiency* page 59; Mrs R. M. Ragg page 83 middle; David Simpson pages 45, 64; Hugh Simson Designs page 43; Smith Kline Animal Health Ltd page 80; Mrs M. Stevens page 83 bottom; Sally Anne Thompson pages 111 top and below, 112 top and below, 113; Mrs V. Thornley page 82; W. M. F. Rabbit Equipment page 168; Wolseley Webb Ltd page 51.

The diagram on page 50 is based on the Forestry Commission leaflet 'Trees and People'. The MAFF extract on page 13 and the table on pages 122–3 are reproduced by courtesy of *Practical Self Sufficiency* magazine.

Line diagrams drawn by G. J. Galsworthy.

CONTENTS

INTRODUCTION

This book is for those who are, or are interested in becoming, part-time farmers, and who may also be interested in expanding their activities on the land, or with livestock, into a small, commercial enterprise. A part-time farmer may well have another job, not necessarily connected with the land. Who are these people? you may ask, if you have not had much contact with what is happening in rural areas. In order to answer that question we need to look at what is happening on a small, but growing scale in the countryside, on both sides of the Atlantic, and in many Western European countries.

An increasing number of people have, in recent years, moved to live in the country on a self-sufficiency basis, providing their own vegetables, eggs, meat and dairy products. At first, they were regarded as freaks and dropouts, and the media, in particular, delighted in portraying such activities as a weird preoccupation with turning back the clock in order to become pre-industrial revolution peasants. Every field of human activity does, of course, have its lunatic fringe, including self sufficiency, but the majority of those in any field are, thankfully, ordinary people who are indistinguishable from anyone else. My experience of editing a self-sufficiency magazine in Britain, and in meeting and talking with active participants in this field in Britain, USA, France, Holland and Germany, indicates that most people have some kind of income that is not directly from the land. Often they are professional or self-employed people, or may have a part-time or seasonal job. Jobs which entail shift work and which therefore allow free time during the day, are also prominent.

In Britain, before the 1950s, people working their own small piece of land would have been called smallholders or cottagers, depending on the amount of land they had, although a smallholder with a profession would have been virtually unknown.

In America, the term used is homesteader, and this was usually a full-time occupation for a family on a relatively small area of land. In recent years, the new homesteaders have increasingly operated on a part-time basis, with the main income coming from a non-land based activity.

Now, we have this new phenomenon – a comparatively large group of people with diverse backgrounds, education and professions, living and working on the land. They defy description because they fit into none of the traditional categories – smallholder, cottager, homesteader – and least of all the rather weak label of hobby farmer, which is sometimes used in British agricultural circles. If there must be a label, then the only appropriate one for such a widely differing group is part-time farmer.

So, part-time farming is increasingly emerging as an agricultural phenomenon, although it must be said that in many parts of Germany and France it has always been a major aspect of commercial farming. In Japan, too, many farmers work on a small scale, operating on a part-time basis, with a high proportion also employed in the car-making and ship-building industries.

Anyone who has read about the remarkable developments in computer micro-circuitry will realize that the implications are phenomenal. Millions of jobs which are currently carried out by human workers are at risk because the technology to replace them is already here. The professions, such as teaching, accounting and the law, will also be radically affected. They have grown around the fact that they are exclusive bodies of specialist knowledge that others do not possess, and cannot interpret. However, with the inevitable spread of information access, even the most specialized knowledge and its interpretation will be available to ordinary people, at the touch of a button. Economists are already predicting that part-time or shared work, amounting to a greatly reduced working week, will become commonplace, and those who have any kind of job will be lucky. The predictions of sociologists are that society will undergo tremendous upheaval in adapting to these conditions, and in particular to the vastly increased amount of leisure.

Part-time farming offers many possibilities to those suited to the way of life. It enables a family or community to provide a large proportion of their own food, it fosters the self-help principle, and it is an excellent and profitable way of making use of enforced leisure. Perhaps, most important of all, it will help to invigorate the life of local villages and communities and provide much needed services on a local basis. It is, for example, absurd that eggs that have been produced in factory farm batteries on one side of the country should be expensively transported hundreds of miles, when people could be eating locally produced and free-range ones. I am not suggesting that it is possible to make a complete living from one small enterprise, but a combination of several income sources can produce a viable and satisfying lifestyle.

This book sets out to show what is possible. A far-fetched idea? Not at all. It is already happening.

How to use this book

It is not essential to read right through the book, although I hope that you will. The first chapter, 'Sorting out priorities', is, however, a fundamental appraisal of basic questions, so it is recommended that you read this. Not all of it will necessarily apply to you, but it does serve to provide a background against which many of the problems of the small, part-time farmer are highlighted. It also covers the difficult questions of finance and marketing.

After this, each chapter is self-contained, with details of a specific topic, such as a particular livestock. The information here is of a practical nature, with advice given on the best methods of management and husbandry.

The final chapter gives ideas and suggestions for other home businesses. As far as possible, existing enterprises have been used as examples, and I am much indebted to the many people who gave up their time to talk to me.

1 SORTING OUT PRIORITIES

Anyone setting out to try to earn a complete living from commercial farming is doomed to failure unless he or she is prepared and able to invest a relatively large amount of money in the enterprise. The part-time farmer is in a different position because, if his prime income source is from some other activity, his reliance on the farming income is less. The established farming community is sometimes scathing in its attitude to the amateur or part-time farmer. What is often overlooked is that many commercial dairy farmers in Wales and the West of England now derive a substantial part of their income from the tourist trade and they are therefore, by definition, part-time farmers. If they had to rely solely on income from milk there would be even fewer dairy farms left in existence than there are now. The really profitable farms are those that operate on a huge, intensive scale, with the highest profitability coming from the arable farms of the eastern counties. One has only to compare, for example, the old tractor driven by the average small Welsh dairy farmer with the sophisticated machinery used by the East Anglian cereal farmer, to have a fair indication of relative profitability.

It is a reasonable assumption to make that no-one would contemplate embarking on a part-time farming venture unless he or she is interested in farming. It is also reasonable to assume that he likes living in the country, which, while it has many advantages over the town, also involves having to do without adequate public transport and many of the services and facilities enjoyed by the urban dweller. One of the greatest advantages of living in the country is that, as long as sufficient land is available, it is possible to produce a high proportion of one's food, and so reduce overall living costs. Is it also possible to earn a living from a small farm? The answer to that does, of course, depend on the individual opinion as to what constitutes a reasonable living. My own view is that it is not possible to make a complete living from a small farm unless there is a high degree of specialization, catering for a high premium market and for this a high capital investment is required. What is not only possible, but is increasingly being demonstrated, is that a small farm can make a valuable contribution in terms of a part-time income, with additional income coming from another activity.

So, it is vital to have an income source before embarking on a part-time farming enterprise. This can be from any one of a thousand occupations, but it is clearly an advantage to have an income from an activity which can be conducted at home. A home business is an ideal situation. But what about the farming side? Many activities are good money savers, but only some will make money. It is important to make a clear distinction between self sufficiency, which will provide for family needs and reduce costs, and a commercial venture which will bring in an income. The question is, which aspects of farming will bring in the best income? This again will vary depending on a number of factors, including individual situation, area of the country, the amount of time spent on the enterprise, the level of capital invested in it, and the amount of time spent in active marketing. It is a question that will be looked at in greater depth in the individual chapters on specific livestock and crops, so that, hopefully, some degree of comparison will be possible.

Opposite: Pigs are reputed to be the most intelligent animals in the farmyard.

An appraisal of basic questions

There are various questions which need careful consideration before starting a part-time farming activity, and it would be useful to look at them in turn.

IS THERE A DEMAND FOR SMALL FARM PRODUCE?

A study of economics would soon indicate that large producers are able to produce goods in greater quantity, more cheaply and with a wider distribution network, than the small producers. What the large producers cannot do as well as the small farmers, is adapt to local demands and cater for specialized needs. So, the small producer immediately has an advantage in his own area, as long as there is a demand to be met. It is here that the importance of thorough research into local needs becomes apparent. Further processing on the farm can increase the value of the product, for example, making cheese and butter from milk.

Another important factor is the growing public awareness of the disturbing practices in large-scale and intensive farming, such as battery egg production, the widespread use of steroids and antibiotics in animal feedstuffs and the treatment of calves in intensive veal production. The level of chemicals, such as inorganic fertilizers, herbicides, fungicides and pesticides used in the production of farm and market garden crops is also of concern to many people. These factors, in conjunction with the current interest in diet and health, have produced a growing demand for wholefoods, unrefined produce which has been produced naturally, without the aid of additives and other pollutants. Experience shows that people with an interest in wholefoods are prepared to pay a higher price for produce they know to be free of additives. It is this specialized market that the small farmer can cater for more effectively than the large producers who are geared to intensive production.

One of the main problems of the small producer is, of course, distribution, and it is vital to have established the adequacy of the market before plunging into production. Production itself is not without its problems, for it is more difficult and more labour intensive to grow produce without the help of the chemical industry. The small, family farm is by its nature labour intensive, although on a small scale, and is therefore in a better position to carry out an enterprise such as this. The question of marketing will be discussed in greater detail later in the chapter.

SHOULD I SPECIALIZE IN ONE ASPECT OF FARMING?

The distinction has already been made between the self-sufficiency approach which caters for personal and family needs, and the small commercial enterprise which is bringing in an income. With the self-sufficiency approach there is often a surplus of produce which can be sold on a local basis, but this is not the same as setting out to produce a marketable commodity. It is, of course, an individual decision as to whether specialization is appropriate. What is certainly true is that it is easier to concentrate on one subject and to do it well than to spread one's effort and energies over a wide variety of activities. The greater the number of activities, the more labour intensive the enterprise becomes. It may be that the main income source itself will dictate that only one farming activity is advisable. There are some aspects of small farming which dovetail together quite well: the keeping of table poultry fits in with free-range egg production, while the plentiful supply of manure from livestock generally, provides the fertility for a specialized organically grown luxury crop such as asparagus. Some activities are purely seasonal; the production of one crop such as raspberries allows plenty of time for other enterprises. Essentially, no-one else can make the decision for you; whether to concentrate on one aspect of farming or diversify, is a personal question.

WHAT DO I ENJOY DOING?

This may seem a flippant question to ask before embarking upon a semi-commercial enterprise, but it is an important one. If you are doing something you enjoy, you are able to bring greater interest and effort to the activity. The best way to carry out an appraisal of what you want to do is to ask yourself some very basic questions. Where part-time farming is concerned, some appropriate questions would be:

- What activities do I find most interesting?
- Do I enjoy being with people?
- Do I like animals?
- Are there any animals which particularly interest me?
- Am I interested in growing things?
- Do I want to be out of doors in all seasons?

WHAT AM I GOOD AT?

This question is closely related to the last one, for if someone is interested in a particular subject, he may already have developed a skill associated with it. Even if no skills have been developed, if the interest is such that the potential is clearly felt and believed in, the learning process and acquisition of experience will follow. Most people are good at something, but again a few questions need to be asked:

- What skills and aptitudes do I have?
- Am I practical with animals?
- Do I have 'green fingers' in relation to plants?
- Am I good in emergencies?
- Am I good at making things?
- Am I mechanically minded?

Different activities need different aptitudes, but only some activities will make money. Where a suitable skill is available, or there is a certain body of experience to be built upon, this provides a good starting point. Once the decision is made, it is relatively easy to establish whether you are as good as you think you are, by going out and putting it to the test. An excellent way of doing this is to spend some time with someone who is already involved in the activity. And as far as organic growing and keeping livestock are concerned, this means spending time on a working farm.

WHERE DO I GET INFORMATION AND EXPERIENCE?

Once a decision is made about which aspect of farming is going to provide a part-time income, it is essential to acquire all the information there is about it. This can come from several sources, but a good starting point is to read all the available books on the particular subject. Magazines are also a useful source of information, not only because their content is usually more up to date and immediate than that of books, but also because they contain information about what tools, equipment and services are available to the prospective small farmer. The better magazines in this field will also place a strong emphasis on readers' personal experiences, and reading these is useful for the realistic picture that they give. Writing to a magazine, saying that you are interested in a particular field, and asking for advice, is also a good way of getting information.

In Britain, the Ministry of Agriculture, Fisheries and Food (MAFF) is a good source of information. They have a large selection of publications, including many free pamphlets, on all aspects of farming. Most of these are geared to the needs of the full-time commercial farmer, but there is much useful information to be gleaned from them by the small part-time farmer. Local MAFF advisers should be consulted because their advice will be geared towards local conditions. The MAFF has a special advisory service called the Agricultural Development and Advisory Service (ADAS). This operates through regional offices and is divided into five sections - the Agricultural Service, the Land Service, the Land Drainage Service, the Agricultural Science Service and the Veterinary Service. Expert advice, information and help on all these aspects is available to farmers who need it. Their advice is free, although certain laboratory tests may have a charge. But as their main aim is to help farmers, not operate a profit-making business, their charges are likely to be much lower than those encountered through dealings with private agricultural consultants.

In the USA, the body to contact for information and advice is the United States Department of Agriculture (USDA).

Talking to people with experience of a particular activity is one of the best ways of acquiring information. Most people are pleased to be asked for their advice and will usually tell you about the drawbacks and difficulties as well as the advantages, and it is especially valuable to hear about the former. Again, the local MAFF office will probably be able to tell you what aspects of farming are being carried out in your area, and may be able to suggest individual farmers to contact. There is also an increasing number of local self sufficiency groups, comprised of small-scale or part-time farmers who have come together on a co-operative basis, and who have a great deal of information on and valuable experience of local conditions to offer.

Lectures and courses on small-scale farming are becoming more readily available and are usually advertised in the small farming magazines. Sometimes, local authorities run appropriate courses, and you should contact them to find out what is planned for the forthcoming adult education activities. If there is sufficient local demand, they can sometimes be persuaded to run a special course. Agricultural colleges often organize special courses on specific subjects, and there are also many privately organized courses on working farms. It is useful to keep an eye on the classified advertising sections of the farming press. In Britain, the Agricultural Training Board and its associated training groups also run courses of a practical nature, which may be open to part-time farmers.

An increasing number of small farms are offering 'working holidays', where tuition is provided in a variety of farming activities. Again, a glance at the classified advertising in the small farming magazines will reveal a large choice, including pig-keeping, home dairying, organic growing, poultry-keeping, and many others. In Britain, a particularly useful organization is Working Weekends on Organic Farms (WWOOF). It puts people living in the town in touch with small farmers who require help with their farming activities at weekends, or for longer periods. The arrangement is that the visitors are given free board and lodging, and tuition in a range of farming and livestock practices, and in return, the farmers receive help in the form of manual labour. This system can work very well, and in some cases visitors subsequently buy their own smallholding to put into practice what they have learnt at first hand.

IS MY AIM REALISTIC IN RELATION TO MY RESOURCES?

If someone is setting out to be a part-time farmer, he must also be part-time something else, and as this other half may well be the main income-earner, it is important to carry out a realistic appraisal of what is involved in starting a farming activity. Time, physical energy and money are all limited commodities. Many part-time farming enterprises are run on the basis of the husband having a full-time job which brings in the main income, while the wife carries out the farming activity and gradually builds it up as a viable enterprise. In some cases, this has been successful to the extent that the husband is able to reduce his job to a part-time one in order to build up the farming enterprise, or even, in a few cases, to giving up his job entirely because it has become financially prudent to do so.

Any proposed farming activity should be subjected to the common sense equation. This is: TEM = CS (Time, Energy, Money = Common sense). In other words, it is only prudent to start an activity if the amount of time, energy and money involved is acceptable to the standards of common sense. This equation works well if common sense is applied, but the trouble with common sense, it has been said, is that it is not very common.

The question of time spent on an activity is an interesting one, because there is so much variation in attitude about what constitutes an acceptable time limit for a particular activity. Many people who work for themselves tend not to cost their time properly, so that the actual income in terms of time spent on an activity is ludicrously small. It is the total time you spend that needs to be assessed, including planning, marketing, paperwork and so on. It is important that you value your time properly, and respect it, because if you do not, no-one else will either. The best way of arriving at a valuation is probably to ask yourself how much an hour you would expect to earn if you were working for someone else. That figure should then be given an important place in future costings, so that pricing can be set accordingly, to give a proper return. Try to match the time to the amount of income required, and remember that a job will take longer than you think. Activities such as making repairs, clearing up, or filling in tax returns all take time and are often not taken into consideration. Productive time, when money is actually being made, is probably less than 50 per cent of the total time spent on a commercial enterprise.

Physical energy is an important factor in the farming world, where many activities require hard work. It is patently absurd for anyone frail, old or infirm to take on an enterprise which might have an adverse effect on his or her health. It is also advisable for someone who has been used to a sedentary job to establish whether he is capable of physical work, and indeed whether he wants to do it. This is where previous experience of working on someone else's farm is so valuable. It is important to plan your activity to work within your current and future energy potential. Evidence shows that self-employed people tend to expend more energy on earning a living than employees do. So, take your measure and fit your activities to suit your potential.

Any enterprise is doomed to failure if the question of money is not regarded seriously. For any enterprise, a certain level of capital is required, and unless you have financial resources, this must come from your income. For this reason, it is not a good idea to start on too big a scale. The best approach is to start in a small way that does not require a large capital outlay, and to develop gradually, investing more money in the enterprise as it becomes necessary and advisable. The question of finance is discussed in more detail later in the chapter.

SHOULD I REGISTER A SMALLHOLDING OR SMALL FARM?

The commercial farmer and smallholder operating from officially designated agricultural land is able to claim grants and subsidies, but in order to be able to do so, he must have an identification number from the Ministry of Agriculture, Fisheries and Food. People buying a farm of a few acres or a house with a large garden and paddock, may find that their land is no longer classified as agricultural land, and they are not therefore in a position to apply for a registered identification number.

If they are able to do so, is there any advantage in making an application? It is impossible to generalize because so much depends upon the scale and type of activities. If the intention is to run a small, commercial herd of Jersey cows, this is obviously in a different category to someone who is selling surplus free-range eggs and a few tomatoes. The best approach is to contact your local divisional office (the address is in the local telephone directory) and talk it over with them.

There is continual confusion over the question of registration, and the MAFF has clarified the situation as follows:

'A smallholding or any agricultural holding, is required to be registered only if the local divisional office of the MAFF, on behalf of the Minister, serves notice to the owner or occupier that the holding should be registered. The notice will take the form of a request for information on area and extent of land held, etc. This registration is required only for the purposes of collecting statistical information. An owner or occupier may make an application himself for registration of a holding to the divisional office but is, in any case, not legally obliged to register his holding unless a notice is served. This form of registration is not to be confused with the issue of identification numbers by divisional offices to holdings receiving grants or subsidies from the Ministry, nor with registration (by the divisional offices) of holdings that are used for more particular farming practices, e.g., dairy farming. These matters require separate application and consideration, and it is the responsibility of the owner or occupier to contact the local authority if he wishes to bring into use any land that was not in use as such.'

So, it is an individual decision as to whether an application should be made, and local advice is best, but do bear in mind that there may be unforeseen consequences. For example, a couple applied because they thought the grants would be useful, but the first thing that happened was that the local water authority sent a large bill, on the grounds that their property was now an officially recognized agricultural smallholding, and therefore merited a higher rate.

WHAT ABOUT REGISTERING A BUSINESS NAME?

The registration of a business name under which you will be trading is not the same as the registration of a farm. If you decide to operate a small business, you can do so under your own name without having to register a business name. If, however, you use a married woman's maiden name or any other name, it will be necessary to register it. The Department of Trade produces a pamphlet entitled 'Notes For Guidance on Registration of Business Names', which is available from the registrar of business names. The address is given in the reference section. All that is necessary is to fill in a form and pay a small fee. The registrar will then tell you if the name is acceptable, and it is worth waiting to receive confirmation before having any cards or stationery printed.

DO I NEED PLANNING PERMISSION?

Even if you own your house, you require planning permission for any activities which are not pertaining to the domestic purposes of the house. Although there are obviously thousands of people who are engaged in activities like selling surplus produce at the gate without planning permission, strictly speaking they require permission to do so. Obviously the question of scale applies here, and taking the common sense approach, it is probably better to start with a very small enterprise, without involving the local authority at the beginning, and see how the venture develops. If it shows signs of rapid development which require the use of a specific building, the local authority should be approached for permission. Establishing a business without permission runs the risk of a fine, and any buildings put up will have to be dismantled. One of the criteria that a local authority applies when considering an application for change of use, is the possible nuisance to neighbours. They are often invited to submit their views before a decision is made. If the decision goes against you, you have the right of appeal.

WILL I NEED TO PAY EXTRA RATES?

If you have applied for, and received, planning permission for change of use for commercial purposes, the planning department will inform the rating department who will probably send a valuation officer to see you. He will arrive, armed with a tape measure and notebook, and will take the appropriate dimensions of the building or room where the business activity, such as a farm shop, takes place. Shortly afterwards, you will receive a rate demand based on the relevant commercial tariff. Generally speaking, you will be expected to pay for this in a lump sum, but the local authority may allow you to pay it in instalments, as is the custom with domestic rates.

WHAT ABOUT INSURANCE?

If you have a mortgage on your property, it is necessary to check with your building society as to whether they have any objections to your operating a small, commercial enterprise on the site. They will not normally object as long as the appropriate planning consent has been obtained from the local authority, and provided there are no restrictive covenants applying to the property. The insurance cover must be thoroughly checked and extended to include the business. If you neglect this and then make a claim on your existing insurance, you may find that the whole thing is invalidated by there being a business activity on the premises. The extent of the insurance cover will depend upon the type of business, the amount and nature of the equipment used and on the number of people visiting your premises. Talk this over with the insurance agent and get advice from as many sources as possible. Your bank manager and accountant will usually have useful advice on this subject and will be able to recommend a reputable insurance company.

WILL I NEED TO PAY TAX?

All income, from whatever source, is taxable, and must be declared for taxation purposes. The advice of an accountant is invaluable, and no enterprise should be set up without consulting one. He will tell you precisely what costs can be offset against tax, including ones that may never occur to you. For example, if you have a business operating from your home, you may be able to claim for the cost of feeding your dog as a business expense on the grounds that he is a guard dog. If your farm is open to the public as a small demonstration centre or farm park, you can claim, as a concession, the costs of buying, feeding and maintaining stock and equipment which might otherwise not be classified as commercial stock. This is most useful, for it means that those who operate an otherwise non-commercial farm may be able to offset a proportion of the costs of stock, feed and equipment as business expenses.

In Britain, if a wife is operating her own business, the husband must pay tax on her profits. He can claim the wife's earned income allowance in addition to a married man's allowance. It is possible to opt for separate taxation but it is only worth doing it if a couple is in a high tax category. Again, an accountant is the best adviser.

NATIONAL INSURANCE

The accountant and social security office will both advise on the question of national insurance. Self-employed people must notify the social security office, and will normally be required to pay Class 2 national insurance contributions, unless gross annual earnings are below a certain figure, in which case they may apply for an exemption. It is important to check whether exemption affects benefits such as retirement pensions. A married woman is normally covered for sickness benefit and retirement pension by her husband's contributions.

A self-employed person is liable to an additional form of tax on annual profits. This is called Class 4 contribution and does not bring any entitlement or benefits. If you are working from home and have an additional job, it is worth reading the leaflet, 'People With More Than One Job', which is obtainable from social security offices. Also, check with the office that you are not overpaying on your national insurance contributions.

VALUE-ADDED TAX

It is necessary to notify the VAT office of HM Customs and Excise if your taxable turnover is likely to exceed a certain figure in the next twelve months. The address of the local office is in the telephone directory.

Once you have registered, you will be given a VAT number, which should appear on your letterhead and invoices. You will be required to charge the appropriate rate (at present 15 per cent) on goods and services where this applies. Some goods are exempt or zero-rated, and food is one of these. The local VAT inspector will advise on what the categories are, and should also tell you how to make out your quarterly VAT returns. VAT on business purchases made by you can be reclaimed and this is usually done at the same time as the returns are sent in. Some people who trade primarily in zero-rated goods are in the happy position of claiming back more than they pay in.

In the USA there is a state sales tax, and it is a good idea to enquire at the local sales tax bureau to see if registration is necessary.

One advantage of having a VAT or state sales tax number (in addition to being able to claim back tax on one's own business purchases) is that it is much easier to buy goods from other suppliers at trade prices. If you have your own business, it is a good general policy to aim to buy as much as possible at a trade or discount price.

SOURCES OF FINANCE

When it comes to financing an enterprise, the wisest course of action is perhaps to start in a small way and gradually expand, so that a large investment is not necessary at the beginning. Most part-time activities will be in this category, and there is a great deal in its favour.

It is necessary to differentiate between farming and non-farming activities, for some of the sources of finances mentioned here are available only to registered commercial farms. The Agricultural Mortgage Corporation will pay 60 per cent of their valuation of what they think a farm is worth to anyone applying for a mortgage to buy a commercial farm. They will also lend long-term capital for the purchase of land and buildings, but not working capital such as for the purchase of stock. It is necessary to provide evidence that the farm will be run competently, and this includes submitting a budget showing how the money will be used, and how it will be repaid. This is not as easy as it sounds, for it will involve doing a Gross Margin Analysis. This is a system of analysing a farm business to see how each enterprise on that farm is contributing to the whole farm. It is necessary to know which enterprises are profitable. With this information it is possible to do a forward projection and estimate the future profitability. The bank manager will help you to fill in the form. The Agricultural Mortgage Corporation is not generally interested in smallholdings, and most of its dealings are with larger farms. A building society may lend money based on the value of the house rather than the land.

Where working capital is required for a farm, there are several sources. Life assurance companies and banks will consider loans. Hire purchase is available for machinery, although the leasing of machinery is becoming increasingly common. The Fatstock Marketing Corporation may lend up to 75 per cent of the money required to purchase livestock, as long as the fattened livestock is sold to the FMC.

Grants and subsidies are available to British farmers, but only for registered commercial farms. This is generally recognized to be where a farm offers full-time employment for at least one person. Part-time farming is not yet given the official recognition it deserves, but, as the number of part-time farming enterprises increases, the situation may change. Details of what grants and subsidies are available can be had from divisional offices of MAFF.

Where a small business is operating from a part-time operation, but is not strictly speaking a farming enterprise, the previously mentioned sources of finance will not apply. However, banks are always ready to consider loans and it is worth talking to the bank manager about your plans for a particular enterprise. The Council for Small Industries in Rural Areas (CoSIRA) is worth approaching for advice and information and possibly a grant, particularly if the enterprise is associated with rural crafts.

KEEPING RECORDS

It is well worth keeping records for any activity, but for a commercial one it is essential. There are basically two areas where records are needed. They are needed for the everyday matters such as the management and welfare of stock, and they are needed for the business itself. The former should include a diary of activities, such as the dates when vermifuges need to be administered, or when calving is due to take place. The latter will include all details of costs, purchases, transactions, sales, profits, VAT and records of everything pertaining to the business.

SOURCES OF LABOUR

This brings us to the question of available help in terms of labour. The small farm-based enterprise will obviously rely heavily on family participation, but there are times when even the tiniest operation will need extra help.

You can either do the secretarial and bookkeeping work yourself, or utilize the part-time services of a specialist. Unless the enterprise is so big as to make it unavoidable, it is best to avoid becoming an employer, as that involves a considerable amount of paperwork and payment of the employer's National Insurance contributions for the employees. It is best to use part-time self-employed specialists who are responsible for their own tax and national insurance contributions.

Local casual labour can be used for many seasonal activities, such as vegetable picking and poultry plucking. The availability of this kind of labour is an important aspect in choosing a farm location. If you employ a full-time agricultural worker there are agreed wage rates laid down by the Agricultural Wages Board. And, as mentioned earlier, you can offer working holidays on your farm.

Marketing

If a commodity is produced, it will need to be marketed, otherwise financial loss will quickly result. Many marketing services are provided for a registered commercial farmer, so that all he needs to concentrate on is the production.

These marketing services include milk, hops, potatoes and wool. The addresses of the appropriate marketing boards are given in the reference section.

CONTRACT GROWING

Contract growing is a system whereby the farmer grows a crop, such as peas, on a contract basis for a company specializing in, for example, food processing. The company will supply the seed and information on its sowing, growing, fertilizing and harvesting, and the farmer carries out the various stages. He will not be able to grow the crop organically, and in fact, has little choice about how to conduct any of the activities. Organically produced apples may be bought by a company specializing in the production of apple juice made from unsprayed apples, but the factory needs to be in an area close enough to make delivery practicable. Cider manufacturers are often interested in local cider apples and may buy your whole crop if it is suitable.

MARKETING CO-OPERATIVES

The Central Council for Agricultural and Horticultural Co-operation will advise, and in appropriate cases, make grants to enable co-operative marketing schemes to be set up. On the whole, they are not as popular in Britain as they are in some parts of Europe, and indeed, some seem to be more effective in the realm of co-operative purchasing of equipment, rather than in marketing. The part-time farmer, by nature of the fact that he is often not taken seriously by the established farming community, is generally left to fend for himself. As a result of this, there has been an interesting development, where local self-sufficiency groups have been formed. These are loose-knit organizations, catering for a wide variety of interests and with the main emphasis on bartering, bulk and discount purchase and mutual help in marketing. For example, in one area, two members have successfully applied for a licence and started an organic butcher's shop which operates alongside a wholefood business. Meat is bought 'on the hoof' from group members, who produce stock on a non-intensive basis, without the use of hormones and steroids. Members' organically produced fruit and vegetables are also sold, and demand exceeds supply.

DIRECT SALES

Direct sales are still the small producer's main outlet, and these will predominantly be to local consumers, retailers, restaurants and hotels. There is no easy way to persuade people to buy your produce. It is simply a matter of going to the shop, restaurant or hotel and showing the proprietor some samples and telling him you can supply him to order. If he agrees, details of delivery and discount can then be discussed. As regards pricing, average prices of a wide range of goods are usually published every week in the farming press. Organically grown produce will generally command a higher price.

FARM-GATE SALES

Farm-gate sales are popular with consumers, but do rely on car access. It will be necessary to provide car parking facilities off the road so as not to create a traffic hazard, and space for turning, so that cars do not have to back out onto a road. It is necessary to have planning permission from the local authority to open a farm shop and if a purpose-built permanent building is put up, this will also need building permission. It is more usual, however, to utilize an existing shed or outbuilding.

Adequate signs are essential and, here again, local authority permission may be needed to erect signs by the side of the road. These should be placed in such a way that they are clearly seen from the road and in plenty of time for the driver to react safely, not with a last-minute swerve. Once off the road, the driver should be welcomed by another clear sign showing him where to park. Once parked, he will need to know where to go and again, there should be a clear indication of where the farm shop is located. There is nothing more annoying than turning in through someone's gate in response to a sign outside and then finding no other signs. 'Closed' and 'Open' signs are also essential, and these should be in a prominent position outside, so that the situation is clear to drivers before they turn off the road.

The farm shop.

In the USA, roadside stands are much more common than they are in Britain, where local authorities generally do not allow direct trading on the side of the road. There are still regulations to be met in the USA, however, and it is worth checking on local, county and state regulations before going to the expense of building a stand.

'PICK-YOUR-OWN'

One of the labour-intensive activities is picking a crop before it can be sold. In recent years there has been a tendency to by-pass this activity by allowing customers to pick their own. This has proved to be immensely popular, particularly at weekends, when family parties go out to pick crops for their freezers, and make the occasion a family picnic outing. It is now so common that there are directories available showing the location of all the 'pick-your-own' farms. In order to be included in these directories, it is usually only necessary to get in touch with the publishers and give them the appropriate details of address, road access, crops available, and times of opening.

The experience of many farmers operating 'pick-your-own' schemes is that it is better to provide standard containers and to sell by weight, rather than by volume. In the case of strawberries, it might be thought that weighing the customers as they go in and before they go out, would indicate a considerable weight of illicitly consumed strawberries. In fact, although there is a certain amount of eating, the truth is that you cannot pick and eat, and pick and keep at the same time. As most people come to pick and keep, the amount consumed is almost negligible. If you do not wish to have dogs roaming around and possibly fouling low-growing crops such as strawberries, it is best to be firm and insist that no dogs be allowed in. If customers do bring dogs, a clear sign in the car park should indicate that they are to be left in the cars. If the public come onto your land, it is best to take out a public liability insurance, but do make it clear to parents that children are their responsibility and that they bring them onto your land at their own risk. Crops which can be picked by the public are generally fruit, such as strawberries, raspberries and plums, and vegetables such as runner beans (pole beans), dwarf beans and sweet corn. Root crops are generally not suitable for 'pick-your-own', although it is becoming increasingly common for farmers to run a potato digger over a field, then provide baskets for hand-picking of exposed potatoes from the surface.

LOCAL MARKETS

Every area has its local farmers' market, where produce is sold, usually by auction, and the sales are arranged by a firm of local auctioneers. In addition to this, there are local markets where a stall space can be rented from the local authority. There is often a waiting list for such a stall, but information can be had from the Markets Inspector of the appropriate local authority. Some people share a stall so that help in the manning of it is available, while self-sufficiency or small co-operative groups often take a stall for the combined produce of all the interested members and the manning is arranged on a rota system. The Women's Institute often has a weekly stall where produce from members is sold. Local events such as fêtes and fairs often welcome traders' stalls and will charge either an agreed percentage of the takings or a set fee.

With the decline in the number of village shops, many villagers are organizing their own community shops once a week or fortnight, in the village hall. This gives the small producer an opportunity to bring his produce along, while providing a much-needed service to local communities.

MOBILE SHOP

Villages and suburbs often have mobile shops which call once a week and sell produce and goods direct at peoples' doors. The petrol increases have made this activity more costly than it used to be, but, where there are sufficient customers, it can still be an effective and profitable way of selling produce. It takes time however, to build up a 'round', and it is not as easy as it sounds. It is also unwise to try to 'poach' on an existing round for this can lead to a lot of trouble from existing traders.

Pick-your-own strawberries.

Farm machinery

A question frequently asked is whether to buy or hire farm machinery. It is a complex question and there is no easy answer, for so much depends upon the type of farming, the scale of activities and the capital available. Some machinery is highly specialized and expensive and is only appropriate to the large farm. It would obviously be inappropriate for the farmer who grows a few acres of arable crops just to feed his livestock to buy a combine harvester. He is better off getting someone else to do it for him on a contract basis. Specialized machinery which is only used occasionally would not generally be bought. Ditching or land drainage equipment is easily available through the auspices of a contractor who will come in and do the whole operation in a short time. Where it is necessary to buy machinery, it is becoming increasingly common to lease it rather than buy it outright. This has obvious advantages for

KUBOTA

Above: A trailer has many uses on a small farm.
Above left: Tractor fitted with a safety roll bar and pulling a small trailer.
Left: One of the range of new small tractors may be more appropriate for the farmer with a small acreage.

those who do not wish to lay out a substantial amount of money for direct purchase. Hire purchase arrangements are available for buying equipment and there are agencies who will lend money for this purpose. Some of these are mentioned in the section on finance.

The type of enterprise will determine what type of equipment and machinery is necessary. Someone specializing in the production of horticultural crops will obviously need a rotovator to till the ground, and irrigation equipment will also be necessary for watering the crops. An organic farmer might find a manure spreader which spreads and disperses natural manure on the land, to be a necessary purchase, while anyone with a relatively large number of livestock would need a tractor with a front end loader for muck clearance. It is not necessary to buy new equipment, for there are considerable sales of second-hand machinery, with some companies specializing in this area. They normally advertise in the farming press.

For a farm of 4 hectares (10 acres) or more a tractor will be useful. It is the general workhorse of the farm and with a few implements can do a wide range of jobs. A trailer is a useful purchase, while attachments such as plough, harrow, seeder and mower will do much of the ground preparation and crop planting. Tractors are expensive but again, are available second-hand. An interesting development in the last few years, has been the appearance of small Japanese tractors which fill the gap which used to exist between the garden-sized tractor and the full-sized agricultural one. Haymaking equipment is relatively inexpensive by comparison with other machinery, and the small farmer with cows or goats may find it worthwhile to purchase this in order to make his own hay. A tractor, cutter, swathe-turner, baler and trailer will be needed.

The following table is a general guide to the different farm machinery, and, hopefully, will help to make the decision about what to buy and what to hire, an easier one.

Farm machinery

Machinery	Main uses	Comment
Tractor	As a general workhorse for tasks too numerous to list.	Probably worth buying for acreage over 4 hectare (10 acres). Smaller sites can probably manage with a rotovator, a hand trailer and if necessary contract labour.
Trailer	An essential accessory to the tractor. Used for carrying everything imaginable.	If you've already invested in a tractor, then a trailer is a foregone conclusion: buy one. If you don't have a tractor, get a trailer to fit your car.
Manure spreader	This specialized trailer has a revolving shredder and blades which chop and spread natural manure onto the land.	Worth having if amounts of natural manure produced warrants it. Particularly useful for relatively large organic farmers. Alternative is to stack manure, let it rot, then spread by hand from a trailer.
Grass-cutter attachment	Pulled by a tractor, this cuts grass for hay. Also useful for clearing rough pastures, nettles and so on.	Relatively inexpensive attachment. Essential for those wishing to make their own hay.
Swathe-turning attachment	Turns hay in order to ensure quick drying.	If hay not turned with this attachment, it will be necessary to turn by hand with a hay rake.
Baler	Bales cut and turned hay.	Makes hay storage easier and more economical of space. (Alternative is to make a traditional stack of loose hay.) Can be done by a contractor.
Plough	Many different kinds of plough available, depending upon type of cultivation needed.	Ask for a demonstration before buying. Only necessary to buy if scale of operation warrants it. Ploughing can be done by an independent contractor.
Harrow	For breaking up clods of earth in previously ploughed land and making soil ready for sowing.	Ploughing and harrowing on a market-garden scale can be done by a rotovator – hired or bought.
Seed drill	A wheeled hopper towed behind a tractor, for distributing seeds through a series of evenly spaced tubes.	Some seed drills are also combined fertilizer applicators.

Only basic farm machinery is included in the table, with priority given to that likely to be used by the small farmer. A visit to the local showroom of a farm machinery company will enable the prospective buyer to see most of what is available. Agricultural shows will also have a great variety on show.

It should not be forgotten that the Land-Rover can be used instead of a tractor, and is capable of towing many tractor implements, but it is more expensive on fuel, although – some would say – safer to drive.

Second-hand dumper trucks, such as those used on building sites, are popular with small farmers, and can be used for a wide range of tasks. Another fact worth remembering is that, for the smaller acreage, a rotovator or hand tractor will usually have a variety of attachments for doing most of the jobs that tractors and implements do on a larger farm.

2 THE FARM

As inflation has increased, so the cost of land has risen, until it has reached the situation in many areas where it is out of the reach of most people, especially first-time buyers. There is, unfortunately, no easy solution to the problem of affording land, although an increasing number of families are pooling their resources in order to buy a shared property. This course of action has positive advantages, but it is also fraught with problems. One of the main difficulties is in getting on with another family, particularly where there are shared facilities. For those who are interested in this approach, there are a number of periodicals which regularly carry advertisements from people wishing to make contact with others who want to buy a shared property. These magazines are listed in the reference section at the back of the book.

Looking at a farm

When a farm is viewed before purchase there are several important considerations to bear in mind. The same questions will help those who already live on a farm, but who are not yet involved in an agricultural pursuit, to examine the potential of their site.

GEOGRAPHICAL LOCATION

The first important consideration is where the farm or smallholding is situated in relation to the nearest town, and what the public transportation is like. In many villages the public transport is almost non-existent, and the availability of shops, schools and services may leave a lot to be desired. Where local sales of farm produce are concerned, easy access to ready markets is essential if a lot of time is not to be spent delivering the produce.

QUALITY OF THE LAND

The quality of land varies from one area to another. It may be heavy clay, quick-draining sand, thin chalky soil, acid peat, or if you are lucky, a friable medium-loam soil. The better soil will grow better crops, and this fact is usually reflected in the comparative land values, with the best agricultural land fetching the highest prices. In Britain, the Ministry of Agriculture publishes land classification maps which indicate the type and quality of soils in the different regions. These are useful particularly to those who are thinking of buying a farm in a specific area. The maps are available from Her Majesty's Stationery Office (HMSO). In the USA, the United States Department of Agriculture should be able to help with information.

In Britain, agricultural land is classified by grades, as follows:

Grade 1	The best quality; suitable for most horticultural and agricultural uses.
Grade 2	Not quite as good, and problems may be encountered with some root crops, such as carrots, but generally suitable for most purposes.
Grade 3	Good quality horticultural crops, such as root crops, may be difficult to grow, but good for grazing and, depending on the climate, satisfactory for cereals.
Grade 4	Suitable mainly for grazing.
Grade 5	Rough grazing only.

CLIMATE AND TOPOGRAPHY

The climate and topography of the land play important roles in its suitability for crops and livestock. Hill farms often have rough grazing and are suitable only for mountain sheep, which are adapted to such conditions. Steep slopes can be inaccessible or highly dangerous for tractors. North-facing slopes can be a problem because of the exposed nature and lack of sun, and land which is waterlogged is the worst of all to do anything with. Prevailing winds can have a major effect on certain crops or young livestock, and checking whether the farm is in an exposed or sheltered position is important. The shape and size of fields are also worth checking up on, to establish how accessible and workable they are. It is a good policy to observe what neighbouring farmers are doing with their land, because, if they are dependent upon it for their livelihoods, you may be certain that they are not growing crops totally unsuited to the environment. The local government office will offer useful advice, and private agricultural consultants also offer a service whereby they look at a piece of land and advise on its condition, and on what crops or livestock would be appropriate. The latter will, of course, charge for this service.

It is a simple matter to establish what the average rainfall and mean temperature is for a particular area. The Meteorological Office or Weather Bureau, as well as civil airports, keep records going back many years, and will provide specific information on request.

THE SOIL FERTILITY

It is not difficult to examine and check the soil in order to establish its structure and pH value. The latter is the degree of acidity or alkalinity, and is crucial to the types of crops that will grow. Potatoes, for example, will become scabbed in soil which has too much lime (too alkaline), and most vegetables will not grow well in very acid soils where lime is in short supply.

A soil auger is a useful tool, as it enables a column of soil to be removed easily for examination. The depth of the top soil (the fertile part) is established by checking the colour of the column. The top soil is dark and the subsoil is lighter. The darker the soil, the more humus it contains. This is the organic part of the soil, derived from decayed plant and animal remains, and provides the fertility.

The ideal soil, when picked up and squeezed between the fingers, should compress into a ball which then readily crumbles, an indication that it is not too heavy and compacted like a clay soil, or too thin and dry like a sandy soil. This is called a loam soil, and is the best type for growing crops.

The degree of acidity or alkalinity of a soil can be established by using a soil testing kit which incorporates a scale from 4.5 to 7.5. This is the pH scale, and a near neutral soil would show a reading of 6.75. Alkaline soils would indicate 7.5, and acid soils, 6 to 4.5. A soil sample is taken and put in a glass test tube or other suitable container, then twice its volume of lime indicator solution is added and allowed to mix well. Strain off the liquid and compare its colour with the colour strips on the scale provided.

In addition to the pH test, it is possible to check the nitrogen, phosphorus and potash levels of the soil, so that any deficiencies in plant foods may be rectified. You can either arrange to have this carried out for you by an agricultural or horticultural consultant, or do it yourself, using a soil testing kit.

DRAINAGE

A soil which is badly drained is a menace. It is easy to establish what the drainage is like by looking at the land: rushes, reeds and patches of water lying on the surface are sure indications of bad drainage. It is possible to improve drainage, as long as the land is not so low lying that it is liable to flooding. The question is whether it is worth it. The cost of draining land effectively can be extremely high, and you would do best to avoid buying land which is waterlogged. If you already have such land, the best thing would be to use it in the best possible way without getting involved in costly drainage exercises. One way of using such land would be to grow trees that are adapted to watery conditions. The willow is a good example of a fast-growing tree which can be coppiced as fuel for the popular woodburning stoves which are common in country areas. Some species can also be coppiced for providing willow wands for the craft market of basketry. If it is absolutely essential to drain land, you should enquire whether any grants are available.

Draining is normally effected in one or more of three ways – ditches, moling or the use of tiles and outfalls. Ditches are also frequently used to mark boundaries and this fact alone makes it important to maintain them in good condition. Where ditches have become filled in, clearing them may radically improve field drainage. It is extremely hard work trying to dig them out manually and the best solution may be to hire a local agricultural contractor to come and do it for you.

Mole drainage

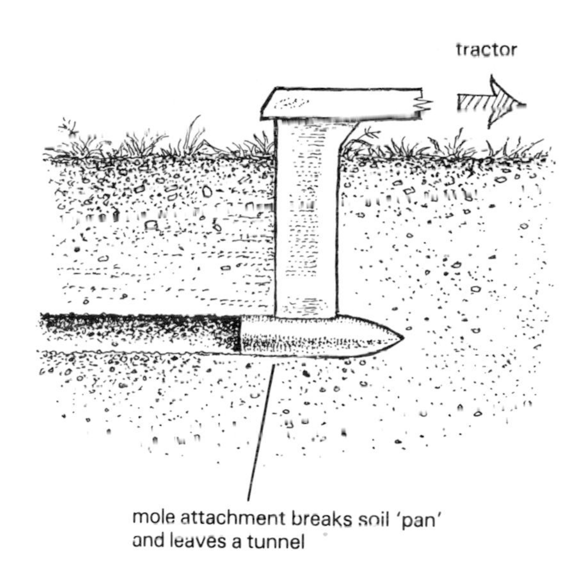

Moling is used in areas where the soil structure has become so compacted that a hard 'pan' is formed on the surface. Water cannot penetrate and so collects on top of the soil. The method for rectifying this is named after the mole plough, which is used to produce a tunnel under the soil surface, so breaking up the hard pan. A suitable clay subsoil is needed; check with an expert. A tractor with a mole plough attachment is used, and it may be more appropriate to call on the services of an agricultural contractor.

The use of tile drains is the most expensive but permanent method of drainage. Here again special equipment is needed to dig the channels, and to place the drain pipes at the appropriate depth and angle for the particular field. These pipes are usually covered over with gravel to assist the water removal before the soil is replaced. An outfall from the drainage system goes into the appropriate ditch. Depending on the type of soil and level of water retention, the tile drains may be arranged in various ways, either parallel or, in severe cases, in a herring-bone pattern, using side as well as main drains. It is essential to have an agricultural contractor to do this job as the cost of the necessary equipment is extremely high. Do check with your local government office to ascertain whether you are eligible for a grant. Small wet areas can be drained by digging ditches for the pipes and replacing the soil.

Tile drainage

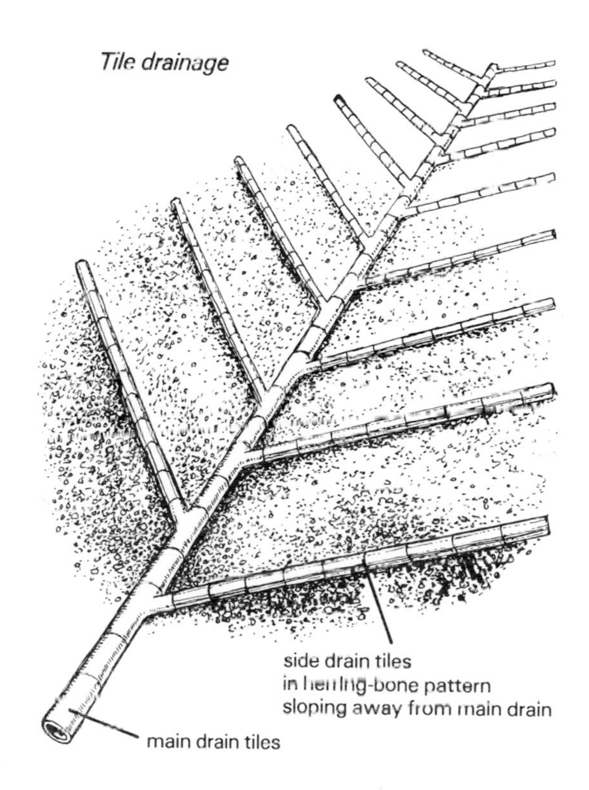

Hedges, walls, fences and gates

Hedges, fences and gates, as well as providing land boundaries, have essential functions in confining livestock and providing shelter. When viewing a farm, it is a good idea to check them carefully because the cost of buying new fencing, or even repairing the old, can be very high. The type of boundary will need to be related to the kind of livestock you intend to keep, or to the crops you intend to grow, should these require wind protection.

HEDGES

Some hedges in certain parts of Britain are hundreds of years old, others are of more recent origin, but a great many have suffered the ravages of modern farming methods. These include 'grubbing out' to create prairie-size fields, making life easier for those who drive large combine harvesters, fire damage from the careless burning of stubble in arable fields, and the killing by toxic agricultural chemicals of many plants and small animals which make up the hedgerow community. The mechanical trimming of hedges, while it is quicker and more economical than manual cutting, also has a damaging and weakening effect on what were once effective barriers to livestock. If a hedge is slashed mechanically, the top growth is cut to the appropriate height, but it does not stimulate growth lower down to make a dense barrier. Rather the effect is to produce gaps through which animals can escape. When cutting by machine, aim for the finished hedge to be A-shaped, tapering to the top.

The traditional method of producing impenetrable and stock-proof hedges was by cutting and layering. This involved cutting part of the way through the main stems of the shrubs

Laying a hedge

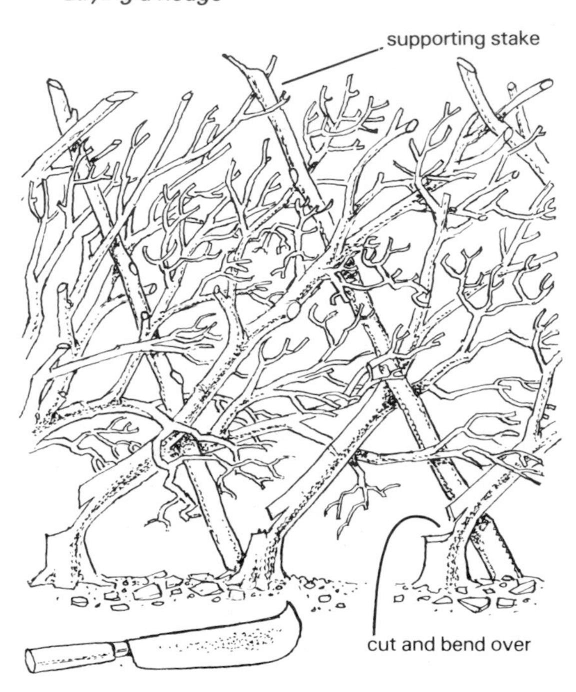

with a billhook and then bending them over to one side. Stakes were hammered into the ground in between the shrubs or small trees and the bent-over growth could be intertwined round or tied to these supports. This layering, when done in the winter, produces new sturdy growth from the bottom upwards, so that the hedge becomes dense and impenetrable. Hawthorn (*Crataegus monogyna* or *Crataegus oxycanthoides*) is the most common hedgerow tree, and responds well to cutting and layering, but

Hedgerow shrubs

Plant	Latin name	Comments
Hawthorn (Quickthorn)	*Crataegus* species	The most common hedgerow plant. Responds well to cutting and layering. Good cover for wild life.
Hazel	*Corylus avellana*	Needs to be regularly cut back or it will grow into a tree. Source of edible nuts.
Elder	*Sambucus nigra*	Grows too open to make a good hedge. Easy to establish and quick-growing, but needs plenty of moisture. Source of flowers and berries for winemaking.
Blackthorn	*Prunus spinosa*	Sharp thorns make an effective deterrent to livestock. Source of sloes for sloe gin.
Dogwood	*Thelycrania sanguinea*	Grows well on chalky soils. Easy to establish.
Rose	*Rosa rugosa* species	Quick-growing and sturdy species. Can be clipped.

A well-cut and layered hedge which will produce new growth from the base.

if planted as a new hedge, the young saplings will need to be protected from stock for at least five years before becoming established. Wire netting put up on a temporary basis will provide protection. It should be added that while most livestock can be confined by well-established hedges, goats will eat even the most prickly of barriers and may need to be tethered.

WALLS

Stone walls, in areas where natural stone is readily available, make superb barriers, but they do need periodic maintenance. It is not a job for the unskilled. Although it may look easy, it should be remembered that the walls have, in the past, been constructed without mortar or cement to hold the stones together. The skill is in selecting suitable stones whose surface irregularities interlock with those of their neighbours.

The walls are constructed by first making a

Construction of a dry stone wall

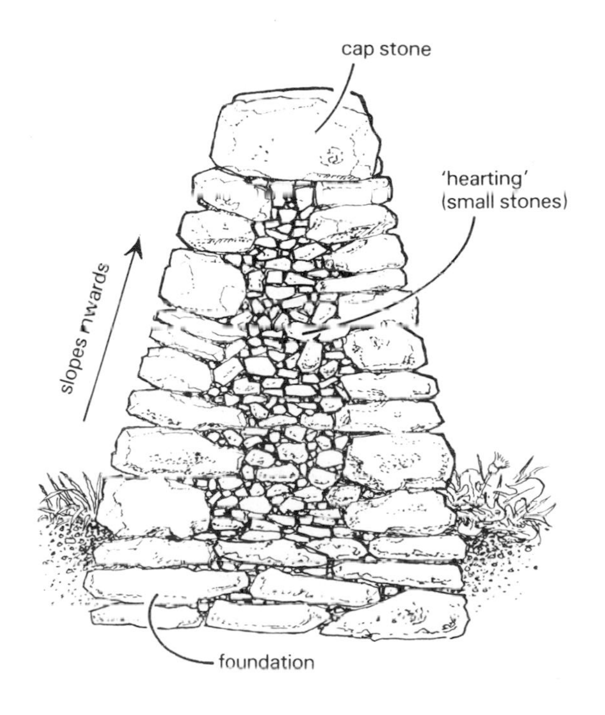

foundation trench about 15 cm (6 in) deep and 15 cm (6 in) wider than the wall. This is packed with hard, fist-size stones and then a double line of string is stretched along the trench, showing where the stones are to be laid. A double row of large, flat stones is then placed along the lines indicated by the string, while the central space is packed with small rough stones called 'hearting'. To provide greater stability, the two sides of the wall usually taper in towards the top.

Once the appropriate height is reached, a cap stone or 'cope' stone is placed on the top. It should be emphasized that this is only one of many methods of dry-stone wall building. In the past, each area had its own distinctive pattern of construction, and the particular wall types are sometimes named after the areas in which they are found.

FENCES AND GATES

Where stone walls or thick hedges are not available, some other fencing will be necessary to contain livestock, or perhaps to keep out trespassers, both biped and quadruped. Post and rail fences are the best type. These have uprights driven into the ground, so that they are secure enough to take horizontal rails nailed into them. The posts are normally 1.8 m × 12.5 cm × 7.5 cm (6 ft × 5 in × 3 in), so that the height is sufficient to allow at least 30 cm (1 ft) to be underground, and the width is enough to allow two rail ends to be butted against each other and nailed on. For horses and dairy cattle three rails up to a height of 95 cm (3 ft 9 in) is normally sufficient, but for frisky bullocks four

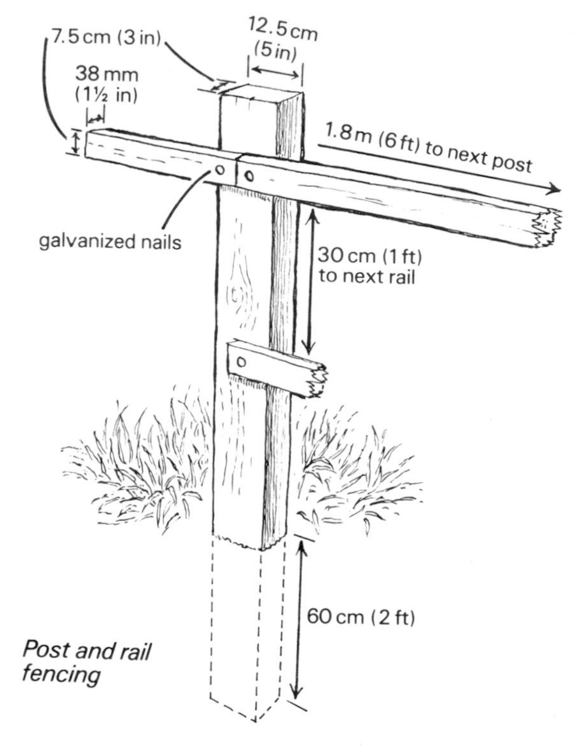

Post and rail fencing

A sturdy post and rail fence makes an effective barrier for most larger livestock.

rails up to a height of 1.25 cm (4 ft 6 in) may be more appropriate. Goatlings, which are generally more difficult to confine than milking goats, may even jump 1.5 m (5 ft) or push their way through gaps in a post and rail fence. It may therefore be more appropriate to use electric fencing for them, otherwise individual tethering will be necessary.

Wire can be used instead of rails, but here the important thing is to strain it properly so that it is taut enough, otherwise sagging will occur. Purpose-made wire strainers are available from agricultural and horticultural suppliers, and it is a good idea to buy or hire one of these if you have a lot of wire fencing to put up.

ELECTRIC FENCING

Electric fencing is often the best solution for those with livestock that is particularly hard to confine, such as goats. It is also useful where controlled grazing is necessary. This allows the fodder to be eaten down completely in one area before the stock is moved on to the next grazing section. Livestock should be trained to respect electric fences by putting food on one side of the fence and the animals on the other side, thus encouraging them to reach under the fence, when they will get a shock.

An electric fence circuit has four parts – the controller or energizing unit which provides the electricity, conductor wires held by insulators, the livestock to be controlled and the ground on which the fence is erected. The controller produces pulses of high voltage electricity, which travel to the fence wire via the output lead. The earth lead of the controller is securely earthed to the ground. The circuit is incomplete until an animal touches the wire and receives a shock as the electric pulse goes through its body to the ground. The energizers are either mains or battery-operated. The former must always be installed indoors and the manufacturer's instructions followed precisely. It goes without saying that all electrical installations should be carried out by a qualified electrician. Battery-operated energizers usually operate either from two 6-volt batteries or one 12-volt battery. Farmers have also successfully used wind-powered generators to energize the fencing system. In Britain, particularly recommended for those who are thinking of installing electric fencing is the MAFF booklet entitled Electric Fencing (HMSO Bulletin 147).

It is impossible to give hard and fast predictions as to what height or type of fencing will confine particular livestock. There will be some sheep which will be content to stay put behind a 90 cm (3 ft) fence, while a Jacob ram may sail over it with impunity. As a general guide, however, the following table indicates what is suitable for the average livestock. One form of fencing which is not recommended is barbed wire, which can inflict cruel wounds, particularly on vulnerable livestock, such as dairy animals.

The electric fence

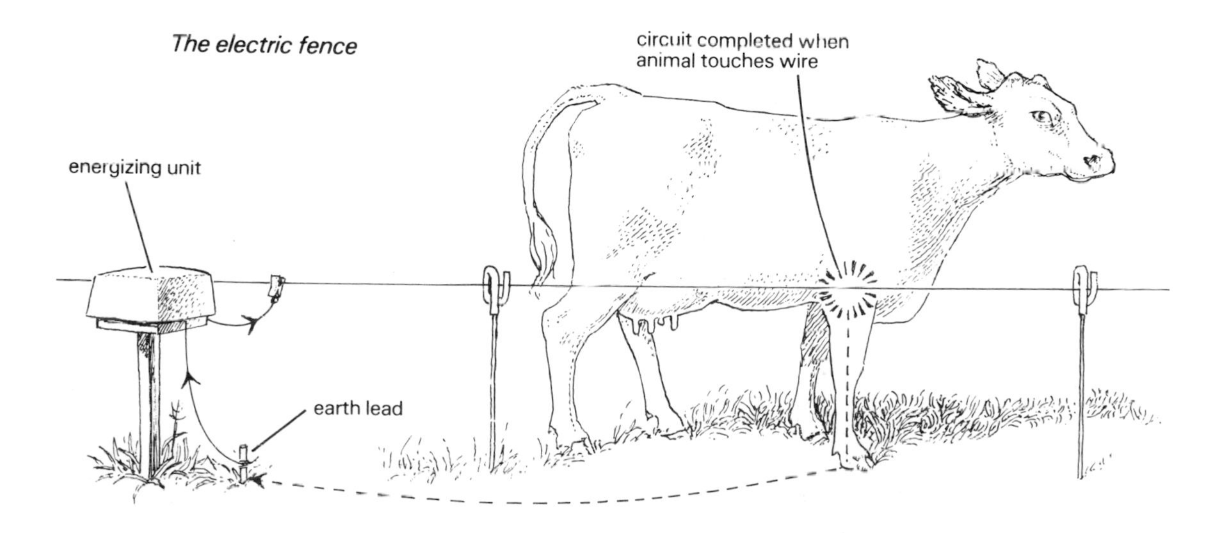

Types of fencing for livestock

Livestock	Suitable fencing	Height	Comments
Chickens	Thick hedge. Stone wall. Poultry netting and posts.	1.5 m (5 ft)	Particularly flighty birds may need to have the tips of the primary feathers of one wing cut.
Ducks	Thick hedge. Stone wall. Poultry netting and posts.	60 cm (2 ft)	Generally easy to confine. The exception is the Muscovy, which may need wing clipping.
Geese	Thick hedge. Stone wall. Poultry netting and posts.	90 cm (3 ft)	Have a tendency to push through gaps rather than trying to get over fencing.
Pigs	Stone walls. Post and pig netting. Electric fencing.	95 cm (3 ft 9 in)	Must be really strong and well anchored, otherwise they will use their noses to push under and lift up the netting.
Sheep	Thick hedges. Stone walls. Post and rails. Electric fencing.	90 cm (3 ft)	Some breeds, such as Jacobs, may need higher fencing or electric fencing, because of their tendency to jump.
Goats	Stone walls. Post and rail fencing. Electric fencing.	At least 1.5 m (5 ft) with no large gaps	Will eat their way through hedges. Milking goats are generally easier to confine than goatlings, which may climb through gaps in rail fencing or jump over a 1.5 m (5 ft) fence. Electric fencing is most effective.
Cows	Thick hedges. Stone walls. Post and rail fencing. Electric fencing for controlled grazing.	95 cm (3 ft 9 in) for dairy cattle, 1.25 m (4 ft 6 in) for bullocks	Bullocks, in particular, have a tendency to lean against fencing and if the posts are weak, may knock it down. Electric fencing is effective.
Horses	Thick hedges. Stone walls. Post and rail fencing.	95 cm (3 ft 9 in)	If you go in for show jumpers, your fences will need to be higher.

Outbuildings

On any farm, the outbuildings are important, because they will house stock, supplies and equipment, or will have a specialized function, such as the provision of a milking parlour. Buildings that are in a poor state of repair may be costly to renovate, although the installation of new buildings is more expensive still. The traditional farm buildings were soundly constructed to provide shelter for livestock, and in the more exposed regions the thickness of the stone wall construction testifies to the importance given to them.

Many working farms today are still utilizing buildings which are over a hundred years old. These traditional structures are lovely in appearance, but they have disadvantages. A notable one is that the access for mechanical muck clearing is often difficult. When they were built, the general access and the doors were on a scale which allowed for mucking out by hand. These days when labour costs are high, even a small dairy or beef farmer might need to use a front-end loader to remove the muck, if he is to keep up with his workload.

For certain purposes, buildings are required to meet statutory regulations. Anyone involved in commercial cow's milk production will need to have buildings which have been inspected and passed as suitable for the hygienic production and treatment of milk. ADAS will advise on the design and layout of buildings.

New buildings may not need planning permission, but will certainly need to meet building and safety regulations. If new buildings are being considered, the planning department of the local authority should be consulted, as well as the local ADAS office who will tell you if you are eligible for a grant. Details of specific buildings are given in the chapters dealing with particular livestock.

A Dutch barn giving weather protection and ventilation is ideal for storing hay and straw.

Roads and access

Traditional farms were often built some distance from the road and access may, in some cases, still be by a track. This can be a real disadvantage, particularly for those intending to take up dairy farming. Milk was, until comparatively recently, transported in metal churns which the farmer put on a collecting platform at the side of the road. From this point the lorry would collect them. Now, all cow's milk is transferred directly into a bulk milk tanker. These large vehicles need to have a concrete road with sufficient turning space before they can gain access to a farm, and this is an important aspect to bear in mind when viewing a farm and its potential. The cost of laying such a private road can be immense. Effective vehicular access is important for other reasons – the delivery of bulk supplies and the transportation of livestock, to mention but two.

When viewing a farm, it is worth checking on an Ordnance Survey map to see whether there are any public footpaths or rights of way. Ramblers and walkers frequently use such paths and, while the majority are undoubtedly careful and respect the farmer's land and crops, there is a minority which is destructive and hazardous. Soft drink cans and bottles and plastic bags thrown away in grazing pasture can cause injury or even death to livestock, while the gate left open is a perennial problem. Some farmers have been forced to erect signs saying 'Beware of the bull' in fields where no bull has trod, because it is the most effective deterrent.

Services

ELECTRICITY SUPPLY

While a farm with no electricity was commonplace a few decades ago, and while paraffin lamps and candles may be romantic, no farm these days can operate without electricity. This may be privately generated in the more remote areas, but the usual source is the mains supply.

Not only the farmhouse but all the outbuildings should have electricity, and this is another important aspect to check. If the wiring is old and out-of-date it may need to be replaced, and power points may need to be installed in specific places. If a lot of machinery is used, it is necessary to have a three-phase electricity supply, and a qualified electrician will need to advise on and install the necessary equipment. The local electricity board will also advise on bulk and off-peak tariffs.

WATER SUPPLY

Water is an essential commodity, and while a private source may be adequate, most farms now receive their supply from the mains source. As they use more water than the domestic consumer, farmers usually have a metered supply so that they are paying for what they actually use. Because of this, it is important to check that the existing pipes and outlets are sound, for if a leak develops in a section of underground pipe, you may be paying for wasted water.

Old farm buildings were not equipped with a water supply and this is an important factor to bear in mind. Can the existing system be extended without too much cost?

Dairy farms must have their water supply tested for purity to ensure that the regulations relating to hygienic milk production are met. Details of specific water supplies, drinkers and troughs are given in the appropriate sections on livestock.

SEWERAGE

Sewerage may be a private or a mains system. With a non-mains system, there is the problem of disposal and the local authority will charge for periodic emptying of the tank. The cost of this varies from one area to another. The main drawback may be that it is not capable of expansion, or that it may turn out to be expensive to expand it.

Pests

Some farming areas have more than their fair share of vermin, including rats, grey squirrels, pigeons, moles, rabbits, mink and even coypu. As far as rats are concerned, it is inevitable that they will be around where livestock is kept or supplies are stored. Local authorities are required to secure the control of rats and mice. If your dwelling and land is not registered as an agricultural holding or farm, you will be regarded as a domestic rate-payer and the local authority will provide you with the services of the Pests Officer free of charge. He will come and inspect your premises and place poison in various places, in such a way that domestic pets will not have access to it. Those living in a registered agricultural holding will be required to pay for the services of the local authority, if they use them. Alternatively, the authority may serve a notice on farmers requiring them to take the necessary action if there is evidence of vermin on their land.

The Ministry's local office will be able to give advice to farmers on the question of pest control, and commercial companies will also have information related to their own products and their use.

A farmer has a statutory obligation to keep his land free from rabbits. They may be shot, gassed or snared, but the use of the metal gin-trap has, quite rightly, been banned for a number of years on humanitarian grounds.

Moles can be a nuisance, particularly where horticultural crops are grown. Strychnine poison is used, but less than it used to be because of the considerable dangers attached to its use. It can now only be obtained by the issuing of permits in approved cases, and an application has to be made to the MAFF Divisional office.

Coypus and mink are both examples of introduced pests, the former being a particular nuisance around the waterways of East Anglia. Farmers are required by law to notify the Ministry if they find either coypus or mink on their land. Advice on control methods is available, as well as the loan of traps. In addition, there are several companies who make traps for sale.

Summary check

The following table is a summary. Take it with you when viewing a farm in order to discover the land's potential. The 'answers' column is left blank so that you can fill in your own findings.

Assessing the potential of a farm

Question	Answer	Question	Answer
Is the house suitable for your needs?		Are there wholefood shops locally?	
What is the state of the outbuildings?		How high is the farm?	
How far is the nearest town and shops?		Is it on steep ground?	
Is effective public transport available?		Is it north or south facing?	
How far are the nearest suitable schools?		Is it sheltered or exposed?	
How far is the nearest doctor/ hospital?		What is the average rainfall?	
How far is the nearest vet?		What is the average drainage like? Are there many rushes and reeds?	
Is there a local market?		Are the fields of a convenient size and shape?	
What grade agricultural land is it?		Is there mains or private water supply? What is its condition?	
What is the pH value of the soil?		Can the water supply be extended?	
What are the prevalent weeds?		Is there a private or public electricity supply?	
What is the condition of the fences and gates?		Is three-phase electricity available for heavy machinery?	
What is the condition of the hedges and ditches?		Is there a local source of casual or part-time labour?	
What trees are present on the site?		Is there evidence of vermin?	
Is there easy vehicular access to a main road?		Is there any game on the land?	
Are there any public footpaths or rights of way across the land?		Is the farm in a tourist area?	
Is there mains or private sewerage? What is its condition?		Looking at all your answers, is the farm really suitable for your needs? (Be honest)	

3 PASTURE & FARM CROPS

Pasture

It has often been said that the most important crop is grass. However, grassland is made up of a collection of grasses, clovers, herbs and weeds. The latter are not particularly desirable unless one has goats, and the higher the proportion of weeds to other plants, the poorer the pasture.

Ideal grasses are those which produce large quantities of leaf growth, which grow rapidly and recover quickly after cutting or grazing. Apart from the older, traditional grasses in long-established pasture, foreign strains from New Zealand and parts of Scandinavia have been introduced, and breeding of new strains takes place continually. Some strains are higher or more erect and are more suitable for cutting for hay or silage. Other strains are close growing or prostrate, and these are generally more suitable for grazing. Strains of particular grasses may also be early or late.

Grasses, like all green plants, carry out an essential function called photosynthesis. This is the manufacture of food in the green areas of the plant using the energy of the sun. Water is drawn up from the soil, via the roots. Carbon dioxide from the air is absorbed through the stomata or 'pores' on the leaves. These substances meet in the areas of the plant containing the green pigment chlorophyll which acts as a catalyst when sunlight falls on the leaves. Almost magically, food is synthesized and stored in the leaves, while oxygen is given out as a waste product. It is this food that provides the most important source of nutrients for grazing animals, and which is captured in hay and silage for winter feeding.

TYPES OF GRASSLAND

There are three types of grassland, depending upon the type of soil and the nature of the plants. The first is uncultivated grassland, which includes hilly and moorland areas of low fertility. This type usually allows only a poor stocking rate, with the emphasis being on sheep breeds that are adapted to poor or hilly conditions. The second is represented by the permanent pastures, which may have been in use as such for hundreds of years. They are never ploughed and have a balance of grasses, herbs and clovers that may be indigenous to that particular area. Many dairy farms have some permanent pastures and they are also associated with a range of different livestock. The third type of grassland is made up of leys or temporarily grassed areas. These may be short leys, lasting up to two years, or long leys, which exist for three years or more. They are part of a general rotation system, and are periodically ploughed up for arable crops. The advantages of leys are that grasses for a specific purpose, such as haymaking, can be grown, growth can be earlier in the year than would otherwise be the case, and there is a disruption to the cycle of parasites that may affect stock. Ploughing in erstwhile leys also helps to build up the overall fertility of the soil.

GRASSLAND PLANTS

Ryegrass is one of the most important grasses. It is a highly productive plant used extensively in permanent pastures and leys, but it does require relatively large amounts of nitrogen. There are two types, the perennial ryegrass, which is available in many different strains, and the Italian, which is used for short-term productive crops, such as hay or silage, or for early grazing.

Timothy is an adaptable grass which grows well under wet conditions and although not early, provides good grazing. It is not as vigorous a grower as some other types, although it has been developed extensively, and there are various strains available.

Meadow fescue is a hardy grass which is often grown with the less vigorous timothy, and is usually found in permanent pasture. It does not grow as rapidly as perennial ryegrass, but it does stand up to considerable treading in by hooves.

Cocksfoot is a tough, rather coarse grass, although it has been made finer by selective breeding. Its great advantage is that it is drought-resistant, and is therefore an important constituent in dry sandy pastures. It is a late grass, sometimes being allowed to grow during the autumn to provide useful winter grazing called 'foggage'.

In addition to these, there are the less commercial grasses, which are nevertheless frequently found in permanent pastures. They often provide good grazing as well as hay or silage, although some are merely taking up otherwise more productive space. Examples are annual meadow grass, sweet vernal grass, creeping bent, crested dogstail, browntop, yorkshire fog, rough-stalk meadow grass, smooth-stalk meadow grass and all the varieties of fescue.

After the grasses, the most important group of plants in a sward are the clovers. These are legumes which have the ability to 'fix' atmospheric nitrogen into the root nodules by the action of bacteria, thereby increasing the overall nitrogen content of the soil. Clovers are also good grazing plants for livestock, providing a useful source of minerals in the diet. They need good drainage and adequate lime and phosphates. There are two types of clover, the red and the white. Red clovers are usually short-lived, lasting up to three years. The broad-leaved red clover is often included with perennial and Italian ryegrass in arable leys. Clover provides useful nitrogen for the subsequent arable crop, and as it recovers well after cutting, it is common to take two cuttings of hay per season of this particular ley crop. The late-flowering red clover is not as vigorous in growth as the broad-leaved one, and is also slower growing. It is usually a component of leys that are left for longer periods. There are many strains of both clovers available.

White clover has a more creeping growth and is useful in keeping out weeds and unproductive grasses which may otherwise become established. They stand up well to grazing and are usually found in permanent pastures and longer leys. White clovers have been much developed and many different strains are available of white, wild white and asilke.

Many herbs are found in permanent pastures, but they may also be sown with grass and clover leys. Particularly useful are burnet, chi-

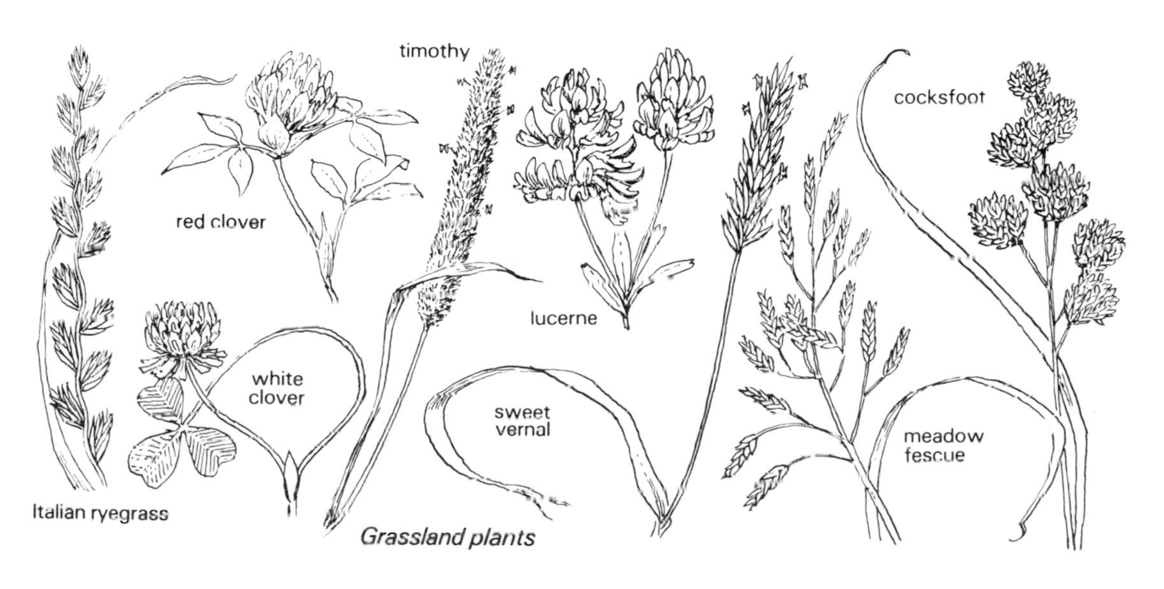

Grassland plants

cory, yarrow and ribgrass. Their advantage is that they are deep-rooted and therefore resistant to drought, and they are a useful source of minerals for grazing stock. The disadvantages are that they may cause heating up in hay and, if eaten in excess by dairy animals, taints in the milk. There are many other common herbs, some of which are relished by stock, while others are poisonous or cause a nuisance because of their invasive nature. Dandelions, sorrel and plantains are eaten by grazing animals, particularly goats, but buttercups and ragwort are poisonous. Make sure that no livestock can gain access to poisonous plants.

MAINTENANCE OF PASTURE

It is out of the question to expect pasture to stay productive all the time unless it is cared for and maintained. Although grass that has been cut or grazed will send up more 'fillers' or leafy shoots than if it were left to go to seed, it cannot continue without being rested and fed. A good starting point is to arrange for a soil analysis to be done, to establish what the soil type and texture is like, what the pH value is, and what the mineral levels are. Details of this are given on page 24.

Applying correctives to land needs to be done at the appropriate time of year. If the land tends to be on the heavy side, there is a danger of tractor and vehicle wheels compacting and damaging the grass or soil structure. Lime, in the form of ground limestone, can be applied at any time of the year, but early to late autumn, when animals are no longer grazing there, is probably the best time. It is best to get a contractor to do this. If you decide to do it yourself, it is important to wear protective clothing, goggles and gloves, as ground limestone is a corrosive substance. It needs to be spread on a dry day, with no wind, and then harrowed in. The amount required will depend upon the acreage to be limed, and upon the pH value of the soil.

Nitrogen, phosphorus and potassium levels can all be corrected by the addition of chemical fertilizers or, in the case of organic farmers, by animal manure or slurry. Again, the time of application depends upon the ground conditions and whether the land is stocked. Modern 'prilled' nitrogen fertilizers can now be applied while animals are grazing. Organic manures and slurries are usually applied in winter if the ground is workable and time must be allowed for them to be incorporated in the ground before livestock is introduced. Organic manures need to be composted under cover for several months before spreading, otherwise 'scorching' of the grass may result. This composting also allows sufficient temperatures to be achieved in

Applying lime.

A harrow attachment.

the heap to kill off parasites, bacteria and some weed seeds. The application of heavy organic manures can be difficult, and it is important to ensure that spreading is even to avoid compaction, leading to bare patches of grass. Efficient harrowing following the initial spreading will ensure this, but for a small mixed farm, an old traditional method of using different livestock in rotation may be appropriate. One method was to allow poultry to follow cattle or sheep so that their scratching did an excellent job of dispersing the droppings, and generally harrowing the soil. The droppings of the poultry are high in nitrogen, phosphorus and potassium and are easily washed in by the rain. An important point to remember is that too much potassium in the soil can 'lock up' the trace element magnesium so that it becomes unavail-

Tractor with manure spreader attachment.

able to grazing animals. Magnesium deficiency leads to a condition known as 'grass staggers' or hypomagnesaemia, which sometimes becomes apparent with new spring grass. Mineral licks and trace element additions in concentrate feeds guard against this and other deficiency conditions.

The best way of controlling grazing is by the use of an electric fence. It is undoubtedly one of the greatest aids of the livestock-keeper, allowing him to manage pasture in the most efficient way. As soon as one area of grass is eaten down, the fence can be moved allowing the livestock to move on. In this way, all the grass is eaten, rather than the more popular grasses being grazed selectively, leaving the coarse ones behind, which is what tends to happen with unrestricted grazing. Further details of electric fences are given on page 29.

If the grass year is taken as starting in Michaelmas or 29th September, which is the traditional starting point, the various activities associated with the maintenance of pasture are easier to understand. Michaelmas was the time when harvesting was complete, the winter quarters of livestock made ready and there was a pause which allowed for planning and preparations for the next year.

The main activities in relation to grassland are detailed in the accompanying table, but it should be taken as a general guide only, for each farm will vary in its activities depending upon weather, nature of land, livestock and the individual farmer.

The grassland year

Time of year	Activity	Comments
Michaelmas – 29th September onwards	Plan for next year. Allocate winter quarters and next year's grazing and hay areas. Apply ground limestone and basic slag if needed. Apply organic manures or slurries if available.	It may be easier to wait until the ground is colder and firmer before spreading manure. Do not apply lime and manure at the same time, as there will be a reaction.
Winter	Check land drains. Clean out ditches. Cut hedges. Repair fences. Complete any jobs not carried out earlier. Plough arable land if a ley is to be sown in the spring.	All the jobs mentioned can either be done yourself if you have the equipment, or they can be done by a contractor.
Early spring	Harrow pasture both ways from early spring onwards. This pulls up dead grass, moss and leaves, breaks crusts and disperses remaining manures. It also dislodges unwanted shallow-rooted weeds and aerates the pasture. Re-sow bare patches. Harrow winter-ploughed land several times to obtain fine soil. Sow new ley and roll straight away.	Again, a contractor can be employed if not doing it yourself. The easiest way of sowing is to use a fertilizer spinner and mix the seed with a little sand to help spread it. The rate is between 7.2–9 kg (16–20 lb) of seed per acre. It may be preferable to sow a new ley in the autumn when it will not normally require bird protection, or it can be sown with a cereal crop. The cereal grows more quickly, and by the time it is harvested, the grass is well established and has meanwhile had protection.
	Apply chemical fertilizers if used. Apply slurry if not applied earlier in the year.	Do not apply more solid manure at this time of year, or grazing may be affected. It should have been spread in winter.
	Roll pasture to level mole hills and to consolidate grasses lifted by frost. Rolling is particularly important on stony ground, otherwise stones may interfere with the blades during hay cutting.	Rolling is not necessary on damp soils, and may not be needed on other soils if the spring is a particularly damp one.
Late spring onwards	Grazing. Control the grazing with an electric fence and watch out for scouring from early lush grass.	

The grassland year cont.

Time of year	Activity	Comments
June	Haymaking. Cut hay as it comes into flower. Turn and bale as soon as possible. Continue to control grazing.	May be better to get a contractor to do this, unless equipment is available.
August	Take a second cutting of hay if ley is good quality.	Alternatively, leave as 'foggage' or late grazing for feeding breeding ewes.
Early autumn	Continue controlled grazing if pasture has not been over-grazed and weather is suitable. Plough land for new ley next season. Harrow, sow and roll the new ley field.	New grass is usually sown before the end of September, to forestall later frost damage. Also, at this time it will not need to be protected from birds.

In autumn, sheep are useful to eat off surplus growth, so preventing frost damage. In winter, animals should usually be left off pasture so that there is plenty of growth at other times of the year. Harrowing lets air into the pasture, and makes it possible for bacteria to break down organic matter, so releasing nutrients. In late spring, scouring can be prevented by putting animals on the grass with hay or straw already in their stomachs.

HAY

Once grass begins to decline in late summer, the nutritional value drops correspondingly, and there is no food value to be gained during the winter. For this reason, grass is cut for storage while it still has nutrients. The hay is available for ruminant livestock, such as cows and goats, so that they can be fed during the winter.

The best time for cutting is just as the grasses come into flower, but this must coincide with a period of dry, sunny weather. In Britain, this is not always easy, which is why close attention to the weather forecast is essential; a standing crop dries out more quickly than a cut one.

The choice needs to be made between taking the hay crop yourself or hiring a contractor to do it. He will cut, turn and bale it and the larger the acreage, the more cost effective this will be. Obviously, for a tiny acreage, the cost per contracted bale could work out the same as that of bought-in hay. Some small farmers have their hay cut by larger farmers who take a proportion of the hay bales in return for their service. Another common bartering system is to repay the larger farmer with one's own labour at harvesting or some other peak activity time. It is important to keep hay dry otherwise moulds causing the condition aspergillosis may appear. This is a serious respiratory condition which can affect a range of livestock and poultry, as well as man himself. The alternative name, farmer's lung, is descriptive of this. Ideally, hay bales are best stored under cover, but with good ventilation; the traditional Dutch barn is ideal.

Tractor with hay baler attachment.

SILAGE

Silage is grass which has been pickled and so preserved as a winter feed. The principle is to cut and chop green grass and press it down tightly in a covered heap from which air is excluded. In these anaerobic conditions, bacteria producing lactic acid get to work, and the acid itself has a 'pickling' effect on the grass. Silage-making is not without its difficulties. It is difficult to make it in small quantities and the bigger the stack, the better the chances are of success, for there is less spoilage. If sufficient grass is available, it is cut with a forage harvester behind a tractor. This cuts the grass and blows it into a trailer pulled by another tractor and kept alongside the harvester. If this equipment is not available, a grass mower such as that used for haymaking will do, but the cut grass will need to be raked up and placed in the trailer by some other means.

The grass is placed in a stack, which may be free-standing, in a pit or in a purpose-built silo. In a stack it is kept covered with a black plastic sheet and after each addition, the tractor is run up over it to press it down and consolidate it. Old rubber tyres or straw bales are then placed on top to keep it weighted down. Silage can be made at any time during the spring and summer months, and does not need to be made all at once. Once the stack is begun, it can be added to periodically, as long as pressing and consolidation takes place after each addition. It is fed to cattle and goats during the winter and can either be cut and given to them in their stalls or in a field, or they can be allowed to feed direct from the stack if it is on concrete. In the latter case, it is essential to control the amount they have access to. The easiest way of achieving this is to use an electic fence, or to place wooden bar or metal rail across the access point.

A forage harvester in action to produce forage for silage-making.

Silage provides an excellent addition to the winter diet, as long as the amount eaten is carefully controlled.

Uses of pasture

How pasture is used will be dependent upon the individual undertaking. For the part-time farmer, hay or silage production will probably only be carried out by him if he has his own animals to feed, for he will not generally wish to sell his hay. The exception is probably those who do not require the hay themselves, but will sell it off the field to another farmer or contractor who will come and cut it for him.

Those with a small acreage and who have a few dairy animals, such as goats, may find it better to keep their own pasture purely for rotational grazing, and to buy in hay from elsewhere. The cheapest way of doing this is to buy 'off the field', from a local farmer. Details of how specific livestock will utilize pasture are given later in the chapter. Information is also to be found in the appropriate chapters on different livestock.

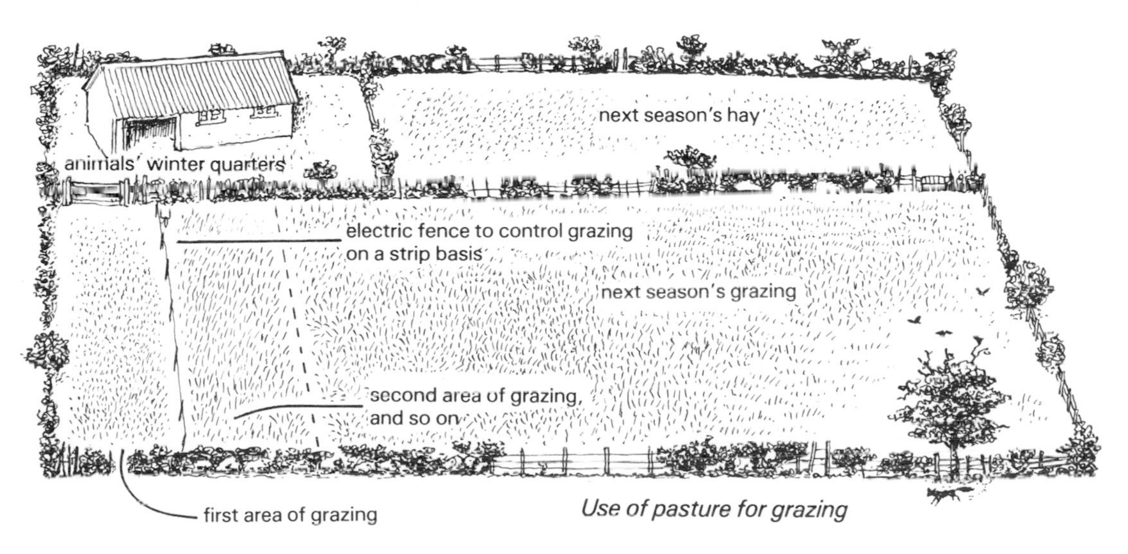

Use of pasture for grazing

RENTING OUT GRAZING

It is common to find fields rented out to other farmers and livestock owners who need some extra grazing. If you are not using the field yourself, it can produce an extra income from being rented to others. For this, it will need to be well fenced, and have water available. If a shelter is also needed and available, this, as well as the water and sound fencing, will make the rental charge higher. If a farmer rents the field on the understanding that the renter is responsible for repairing the fences, then obviously a lower rent will need to be charged.

There are several points to remember if you are considering letting some of your grazing land. The first is that the livestock should be healthy and, where necessary, should have the appropriate 'movement order'. Insist that the animals are wormed before they are introduced onto your land, and ask to see the necessary documents relating to their movement.

Secondly, ensure that the ground is not overstocked, and that you know how long the animals will be there. The table below gives an indication of stocking densities. Thirdly, make it clear that the land is let on the basis that you will accept no responsibility for the livestock. This is important in case of an accident, either to the livestock themselves or to the public in the event of their escape. It is a good idea to have a word with a solicitor before going ahead.

RENTING OUT TO CAMPERS

In the tourist areas of the country, there is a demand for camping space, and this can produce an added income for those who have suitable fields. It is necessary to have permission from the local authority and it is wise to have an initial discussion with them. The local tourist office will also give advice and information, and will put you in touch with camping organizations who are always looking for available camping space. Organizations catering for members with caravans and motor caravans are particularly interested in making contact with farmers. As long as no more than five motor caravans are parked at any one time it may not be necessary to have permission, but local authorities vary in different areas.

It is normal when letting out land for camping, to provide piped water, toilet and litter facilities and level ground. Good access from the road is essential, as well as adequate turning space. It is wise to take out a public liability insurance cover if members of the public use your land.

GRAZING YOUR OWN ANIMALS

Between late March and early October, grass will be available for grazing, with the peak time being between April and late June. The choice of grazing animal will, of course, be dependent upon the individual farmer.

Pasture requirements

Livestock	Type of pasture	Stocking density	Comments
Dairy cows	Good quality permanent pasture or ley.	5 per hectare ($2\frac{1}{2}$ acres)	This density is only possible on first-rate grass, and by using electric fencing for controlled paddock grazing. Poorer pasture will support 1 cow per hectare ($2\frac{1}{2}$ acre).
Dairy goats	Rough pasture and scrubland.	10 per hectare ($2\frac{1}{2}$ acres).	Goats tend to browse rather than graze, and will do a good job of clearing broad-leaved plants and weeds from pasture and scrubland.
Beef cattle	Average quality pasture and leys.	5 per hectare ($2\frac{1}{2}$ acres).	Fences must be strong to contain large bullocks.
Sheep	Anything from rough moorland pasture to good leys, depending upon type of sheep.	10 per hectare (1 acre) if good quality grass. The poorer the grass, the less sheep it will support.	Some sheep, notably the hill breeds and the more primitive breeds, will eat heather and willow bark if available.

Pasture requirements cont.

Livestock	Type of pasture	Stocking density	Comments
Pigs	Average pasture as long as it is not waterlogged	20–25 per hectare (2½ acre) depending on size.	Over wintering breeding sows will do a good job of cultivating and manuring small fields or enclosures ready for spring planting. If you want to keep the grass, the pigs will need ringing.
Geese	Reasonably good pasture. Needs frequent 'topping' to make short young grass available.	250 goslings per hectare (2½ acres) from late April to early Sept. only. 22–25 adult breeding geese.	Useful as followers of larger stock which crop the long grasses down to an available size.
Ducks	Poor to average pasture can be made use of.	500 young ducks per hectare (2½ acres) for one season only.	Useful as slug and pest clearers from strawberry fields before fruit formation and after harvesting.
Chickens	Grass previously cropped by cows or sheep. Shorter grasses preferred to longer, coarser ones.	625 per hectare (2½ acre).	Also useful to turn onto harvested arable fields to glean remaining seeds. Will clear residual parasites from fields vacated by larger stock.

More specific details are given in the sections dealing with the appropriate livestock. It needs to be emphasized that the table should be taken as a general guide only, for there is a wide variation in quality of pasture. Unless otherwise indicated, the stocking densities assume temporary occupation for one season while the grass is available.

Crops

The growing of crops will be largely determined by the scale of operations, by the time, equipment and capital available and by the purpose for which they are envisaged.

The distinction needs to be made between farm crops grown on a field scale, and horticultural produce from a market garden operation. The former will require a high degree of mechanization and a decision on whether to buy machinery or hire contract labour will need to be made. Where crops are grown on a garden scale, there is a further distinction which needs to be made; whether the vegetables or fruit are being grown to cater for family needs or as commercial crops. There is a considerable saving to be made by the growing of one's own fruit and vegetables, and by selling or bartering the surplus, but this is not the same as aiming for a commercial crop.

The first priority in the selection of a commercial crop is to choose one that does consistently well on your particular soil. In this respect, ADAS will advise on the choice of seed varieties for particular soil and climatic conditions, and the local MAFF office will advise on suitable crops. The National Institute of Agricultural Botany produces recommended lists of varieties, which will also help in the selection process. Local farmers, market gardeners and members of local horticultural organizations are worth consulting, for they will have on-the-spot knowledge of local conditions and suitable varieties.

The second priority is to establish what there is a need for. There is no point in concentrating on a suitable crop for the soil, if there is no demand for it when harvested. It is here that your original market research is so important. There are certain crops which can only be marketed through specific agencies. For example, under the Hops Marketing Scheme 1932, hops may only be sold to or through the Hops Marketing Board and growers must be registered with them. Here again, the local MAFF will provide invaluable advice and information.

Organically grown crops

Organically grown crops are those which have been produced by more traditional methods, without the use of artificial fertilizers and pesticide sprays. There is a small, but growing demand for such crops, and the situation at present is that demand exceeds supply. It is here that the small grower or part-time farmer can not only compete with, but beat the large farmer and grower. There are plenty of people growing commercial crops with the help of chemicals but comparatively few producing organically grown quality food.

The Soil Association has helped organic growers and consumers by drawing up a list of standards in relation to soil husbandry and cultivation methods. Any grower can apply for recognition by the Soil Association, and if, after inspection, his practices are acceptable, he is then allowed to display the Soil Association symbol of organic quality when he sells his produce.

The Organic Growers Association is a new organization which seeks to provide information pertinent to the needs of organic growers. It organizes conferences on various aspects of this activity and gives useful advice on techniques and on marketing.

The Organic Farmers and Growers Ltd, is a marketing organization which assists its members with sales and distribution. They have two standards of produce, referred to as OF1 and OF2. The former is the same as the Soil Association standards, while the latter allows the use of fifty units of nitrogen, thirty units of phosphate and thirty units of potash per acre, per season. OF2 is mainly for farmers who are turning over to organic methods, or whose soil is not yet good enough for organic growing. In the USA, the Rodale Organization plays a valuable role in providing encouragement and information about organic growing.

A seed drill attachment.

Farm crops

Farm crops are those which are grown on a field scale, and for which mechanized cultivation and harvesting equipment is needed. They are generally grown as a source of food for the farmers' own livestock, on a contract basis for a concern such as a food processing company, or as a cash crop which the farmer will either sell direct to a retailer or agent, or through a marketing board.

The equipment needed for cultivation such as this will be a tractor, plough, harrow, fertilizer sprayer or manure spreader, seed drill trailer and harvester. The latter will depend upon the particular crop. A potato harvester, for example, may include a combined digger and elevator which delivers the crop through a chute into a trailer to the side of and just behind the tractor. Many farmers, however, still rely on harvesting potatoes by going over the field with a 'digger' to expose the potatoes, then hiring casual labour to pick them up and put them into baskets.

Plough attachment.

CEREAL CROPS

It is unlikely that the small part-time farmer will go in for cereal crops unless he already has the necessary equipment for harvesting them. A modern combine harvester would be economically viable only for a large arable enterprise. It is still possible to buy older second-hand combines, as well as binders, which will cut the crop complete with long stalks, so that bundles or 'stooks' can be made prior to threshing or removing the 'ears' of grain. Thatchers need long stalks and will often come to an arrangement with a farmer whereby a certain acreage is grown just for thatch. The thatcher will either get the farmer to cut the field for him or will arrange for a contractor to cut it with a binder. It is possible to arrange a contract with a thatcher whereby you merely rent the land to him, and he gets a contractor to do everything else including the sowing and harvesting. He will not normally want the grain, so once it has been threshed, you could take it as part-payment for the land rental. This is a painless way of having your own arable crop.

Another way of having arable crops on a small scale is to plough or rotovate the area yourself, broadcast the seed by hand, and then cut it with a scythe or the hay-cutting attachment of a rotovator. The crop will still need to be stacked and threshed to separate the ears of grain and the whole task is extremely time-consuming. It may be more appropriate to get a contractor to do it for you, as long as the acreage is sufficient to make it economically worth doing. It is sometimes possible to get a local farmer to help, and to come to some arrangement with him. Organically grown cereals may be worth growing on a contract basis, for suppliers of stone-ground flour, and in this case, it may be better to hire a contractor to harvest it for you.

It should be remembered that wheat and barley grow much more successfully in the south-eastern areas of Britain.

Winter cereals are the highest yielders and also have the highest value in terms of income per acre. Oats grow in cooler areas, and there is a ready market for it as a feed for horses, as well as a winter feed source for a number of animals and poultry that are being reared on a non-intensive basis.

OTHER CROPS

Sugar beet is an important crop, but needs to be grown in the areas where there is a beet processing plant for it. It must be grown under contract to the British Sugar Corporation. Once harvested, it is transported to the nearest plant to be processed into sugar. An important by-product is pulp which is dried and sold as food for cows, horses, sheep and goats. They will also eat the wilted tops of the plants if these are cut and made available to them in the field. The tops can also be used for silage.

If a farmer grows more than 0.4 hectare (1 acre) of potatoes, he must be registered with the Potato Marketing Board if he intends to sell his crop. The Board grants a basic area quota which determines how much area of potatoes the farmer can plant for the next season. A guaranteed price is given for the potatoes when sold, and if, for any reason, this falls below the predicted level, the government will make a deficiency payment. Potato growers may need specialized planting, harvesting and irrigation equipment, and it is not generally an activity that the small farmer will be involved with. He will probably confine his potato growing to producing enough for his own needs, as well as a certain amount for feeding livestock.

Mangolds, fodder beet, swedes and turnips are important fodder crops, although they are dependent upon the type of soil and geographical location. Mangolds and swedes have been largely replaced by fodder beet in recent years.

Turnips and swedes are sometimes grown by sheep farmers. The flock grazes on the crops in the field, and grazing is controlled by electric fencing. The turnips which are more susceptible to frost, are grazed first, followed by the more hardy swedes. On heavy soils and where winters are severe, the swedes are harvested and stored in clamps for winter feeding.

Other forage crops for livestock feeding are kale, rape, cabbages, lucerne and cereal-legume mixtures. All these crops can either be grazed in the field or cut for silage. Further details of forage crops are given in the table.

Fodder crops

Crop	Cultivation	Comments
Carrots	Sow in fine seed bed without fresh manure, which would cause forking. Thin out to allow room for growth. Store in sand after harvesting.	Not practicable on a field scale. Worth growing in small quantities for a few livestock, such as goats and rabbits.
Cow cabbage	Grow like ordinary cabbages. Can follow early potatoes. Either harvest as needed and feed in stall, or graze in field.	Where grazed in field by cattle, may become mud-splattered because it is low-growing. May be better to concentrate on kale.
Field beans (tic beans)	Winter-sown crop will give heavier yield and will have fewer aphid problems. Hoe between rows. Harvest when leaves begin to dry.	Ground or kibbled beans mixed with other food will provide useful protein for cattle, sheep, pigs and poultry.
Fodder beet	Sow in good seed bed in April. Keep hoed and harvest before frost. Store in a clamp.	Wilted tops and roots good for cattle, pigs, sheep and goats, but do not give billy goats too much.
Jerusalem artichokes	Plant in late winter when ground is suitable.	Not grown on a field scale, and better kept to small areas. Pigs can graze and dig up the tubers direct from the soil. Goats and poultry will eat the cut-up tubers.
Kale	Sow April–May 1.35 kg (3 lb) of seed per acre. Thin out and keep hoed. Can be strip-grazed in field or cut and fed in stall.	An excellent source of winter food for cattle, sheep, pigs, goats and poultry. Beware of over-feeding, leading to 'bloat' in ruminants.
Mangolds	Sow in April in good seed bed, previously ploughed and worked. 4–4.5 kg (9–10 lb) of seed per acre. Keep hoed and weeded. Lift before frosts and store in clamps.	Do not feed until January when they are fully matured, as they are slightly toxic before then. Suitable for most ruminant livestock.

Fodder crops cont.

Crop	Cultivation	Comments
Rape	Sow in spring for grazing in late summer, or after a cereal crop when it will provide late grazing.	Suitable for cattle and sheep.
Swedes	Sow May–June in a fine seedbed at a rate of 450 g (1 lb) of seed per acre. Thin out or single. Harvest in winter and clamp or allow sheep to graze on land direct.	Suitable for sheep, goats and cattle.
Turnips	Sow June–July in a fine seed bed, and treat in the same way as swedes.	Suitable for sheep, goats and cattle.

Trees

Planting trees has an added dimension over the planting of crops generally. Anyone who has ever been involved with it will know that there is a feeling of 'planting for the future'. The timelessness of trees reminds us perhaps of a larger scale of being.

There are many places where trees can be usefully planted, and which are of no use for any other purpose. These include damp, low-lying ground, hilly areas, awkward corners of fields where machinery cannot gain access, field perimeters and boundaries and stony or infertile ground generally. Many of our poor devastated hedges could also be rejuvenated by the judicious planting of trees. It should not be thought, however, that any tree will grow in any old poor soil. It is important to select trees for particular areas, and to maintain them properly. It is estimated that half the trees that were planted in Britain in the 1973 'Plant a Tree in '73' campaign died from being planted in unsuitable conditions or from lack of proper care.

Trees can be planted for many reasons — as an investment crop for felling, for coppicing for a particular purpose (the upper part of the tree is cut down, leaving a stump from which new shoots grow, for subsequent cutting), or for generally improving the landscape, usually referred to as 'amenity' planting. In Britain, mixed woodland yields about 1525 kg ($1\frac{1}{2}$ ton) of coppiced wood a year, but with a species such as the South American beech (*Nothofagus procera*), the yield can be as high as 5080 kg (5 ton) of coppicings a year. It is estimated that the South American beech will be ready for harvesting between twenty-six and twenty-eight years after planting.

Christmas trees are another source of income. The advantage with these is that they can be sold while still fairly small, for the average family tends to buy a tree which is no more than 90 cm (3 ft) high. Christmas tree enterprises tend to fall into one of two categories: the large-scale operation that is catering for the retail trade and supplying shops with trees for resale, or the smaller 'choose-your-own' operation. The latter concentrates on local customers who come to view the section of plantation ready for harvesting, and reserve their trees. They are then tagged with the name and address of the customers, who come to collect them just before Christmas. The advantage of this system is that there are no transport costs, but it is of course necessary to provide adequate car parking facilities. One such small enterprise even allows customers to dig up their own trees, so that they can be sure of getting the whole root system, in case they want to plant the trees outside after Christmas.

The Christmas trade also calls for holly and mistletoe and where these are available, it is worth harvesting them either to offer to shops interested in selling them, or to sell direct to callers. It is not worth trying to grow your own holly from seed other than for interest's sake because it is extremely slow growing. Where an existing hedge occurs it is certainly worth exploiting, as long as cutting does not become excessive and so damage the trees. Mistletoe is difficult to establish, although it is worth a try, and is often found in orchards, growing on apple trees. It does, however, grow on a variety of different trees in addition to fruit trees. It is a partial parasite in that it sends suckers into the sap system of the host tree, but being a green plant, it also produces a proportion of its

Logs need effective support when a chain saw is used.

own food by photosynthesis. There is no evidence that mistletoe damages the tree on which it grows, as long as it does not become excessive.

If mistletoe and holly branches are cut some time before Christmas, it is best to store them outside under netting and with the ends plunged in coarse sand. In this way, they will keep fresh and the berries will not fall off or become prey to winter-hungry birds.

In Britain in the last five years, there has been a booming interest in woodburning heaters and cookers. These appliances burn wood efficiently and provide heat at a fraction of the cost of oil, gas or electricity. The wood, however, needs to be a rural and local resource, because it is needed in large quantities and transport is expensive. People in rural areas who have equipped their houses with woodburning appliances are always interested in local wood supplies, whether this be a delivered commodity or available for collection by callers.

Where there is existing woodland on a farm, it is worth looking into the possibilities of coppicing some of the growth and selling it as fuel to local customers with wood burners. Where there is insufficient to provide a surplus, it is probably worth doing it in order to provide an alternative and cheaper fuel source for one's own use. Some farmers are already doing this, as well as utilizing surplus straw to fuel large straw burning appliances.

Power equipment is necessary, and for pruning and coppicing trees there is no efficient alternative to the chain saw. Doing it by hand may be the traditional and possibly safer way, but it is certainly much slower. Once cut, the branches will need to be cut into log size, and again, a chain saw, with a saw horse to hold the wood, will do this in a short time. It should be emphasized that chain saws can be lethal in inexperienced hands, and basic training to cover the safety aspects is vital. Several agricultural colleges now offer short courses on arboriculture and tree surgery, which include practical training in the use of chain saws. The Forestry Commission also produces much informative literature to help those with an interest in forestry and timber products.

Clothing is an important factor in the safety aspect of using chain saws. Light-weight clothing and a zippered jacket to provide freedom of movement without stray ends is essential, and trousers should be tucked into good, protective boots. A safety helmet, earplugs and gloves with a good grip are advisable, and if the wood is likely to splinter, goggles are a good idea. When buying a chain saw, it is wise to deal with a reputable firm who sell saws with an established brand name. A chain saw equipped with an automatic chain brake will protect the user from a potentially dangerous 'kick-back', by stopping the cutting chain in 6/100th of a second. Large logs may require splitting before they are suitable for use in woodburning appliances, or for selling to customers. Hand log splitters are now available, but while many of them are extremely efficient for the production of one's own firewood supply, a powered log splitter would be needed by anyone involved on a larger scale. There are several models now available, which are used in conjunction with a tractor, and which therefore dovetail well with an existing farming enterprise. Once cut, the logs should be stacked and covered so that they have an opportunity to dry out. The drier the wood, the less tar will be deposited during burning.

Some part-time farmers, in addition to selling wood supplies, have also set up small businesses selling woodstoves, associated equipment such as chimney cleaning equipment and books on the subject. Anyone interested in this aspect, is advised to contact the trade organization Woodburning Association of Retailers & Manufacturers, WARM, which represents the interests of those who sell woodburners in Britain. In the United States the Wood Energy Institute is the organization for manufacturers, distributors and dealers, while the Canadian Wood Energy Institute serves the interests of those in Canada. The addresses are given in the reference section.

If the ground is low-lying and damp, willows and poplars are the best trees to grow. Poplar is not traditionally a good burning wood for open grates, although it is perfectly good in wood stoves. It is a popular choice of those who wish to plant trees as a long-term investment crop, which is felled for timber. In this case, the wood is primarily used by the building and carpentry trades.

The cricket bat willow (*Salix Alba Coerulea*) is another worthwhile investment crop where the ground is damp and unsuited for other agricultural uses. It is possible to buy young saplings from specialist suppliers who will then buy back the timber crop when it is felled and use it for the manufacture of cricket bats.

The variety *Salix vimnalis* is the willow which is used in basketry and the cuttings or coppicings from this is often used for basket making. Specialist craft suppliers, as well as sections of the domestic furnishings industry are interested to buy such willow osiers on a contract basis. Where a plantation of trees is to be planted for timber production, the Forestry Commission will provide a great deal of useful advice and information, covering all aspects of planting, maintenance and harvesting. In addition, there are voluntary organizations such as the Tree Council and The Men of the Trees, who promote the ecological and environmental aspects of tree planting.

Grants are available from a number of sources. The Ministry of Agriculture, Fisheries and Food awards grants for the planting of shelterbelts on farms and horticultural holdings in England and Wales. In Scotland it is the Department of Agriculture and Fisheries which is the body to approach. The Countryside Commission provides grants to local authorities for amenity tree planting in the countryside of England and Wales. Local authorities will, in turn, provide free trees and sometimes planting aids such as posts and tree-ties to those within their areas of jurisdiction who wish to plant trees. This is on the condition that the trees represent a public amenity, that they can be seen from the road and improve the landscape. Most district councils and county councils have a Forestry Officer who will ensure that those who want help receive sound, professional advice.

The Forestry Commission will give grants in respect of areas of tree plantings of 0.25 hectares ($\frac{1}{2}$ acre) or more, as long as the aim is timber production. Both conifers and broadleaved trees are eligible, but the broadleaved ones attract a higher rate. Applicants will be required to work to a five-year plan approved by the Forestry Commission.

TREES FROM SEED AND CUTTINGS

Growing trees from seed is easier than many people realize, and there are several seed companies which now make available a wide range of varieties. The seeds can be sown in seed compost, but in containers which are taller than normal, to allow for the rapid growth of the root system. Once germinated and growing strongly, the containers are put outside in a nursery bed. This could be a layer of sand or peat on which the pots rest, while the whole is protected with netting if rabbits or pigeons are a pest.

Most trees are hardy and the saplings will require little else in the way of looking after, apart from regular watering, and possibly shading in the event of particularly hot sun. When they are about 45–60 cm (1 ft 6 in–2 ft) high, they are ready for planting in their permanent positions. Some people have found that they can sell their surplus tree seedlings quite easily, particularly to garden centres or to those interested in bonsai cultivation.

It is often possible to find self-sown tree seedlings such as oak, chestnut and maple, growing in woodlands, and these can be uprooted carefully and transplanted for growing on. Cuttings can also be taken quite easily from willows and poplars. Where the ground is damp, cut wands or branches from these two species can be hammered into the ground, and will root within a matter of weeks. It goes without saying that no self-sown seedlings or cuttings should be removed from private woodlands without the permission of the owner.

Planting a tree

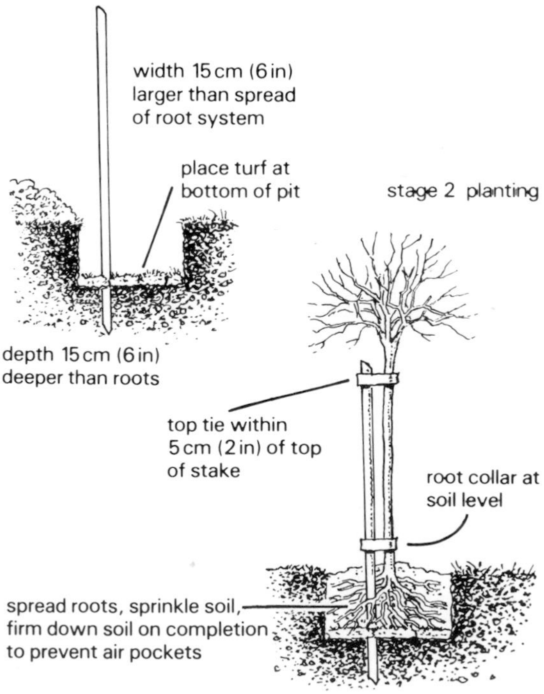

TREE PLANTING

The best time to plant trees is in the autumn or winter, as long as the ground is frost-free. Dig a hole deep enough and wide enough to accommodate the root system, and break up the soil at the base of the pit, incorporating well-rotted farmyard manure or compost. Place the tree in the hole, spread out the roots, and replace the soil firmly. Hammer in a stake to support the tree and attach it with a tree-tie which holds the tree firmly, but without damaging the bark. Purpose-made tree-ties and spacers are available commercially, but a good alternative is to use old nylon tights or stockings. Newly planted trees may also require protection against the depradations of rabbits and deer, and it is here that wire mesh guards are useful. Again, these are obtainable commercially or may be made by putting up posts and wire netting.

COPPICING

Some trees such as poplars and South American beeches are grown for timber production and then felled when they are big enough. Other trees are cut leaving a stump so that new shoots are produced which grow upwards and are then cut for a particular reason. The willow (*Salix vimnalis*) produces shoots or osiers for basketry. Hazel (*Corylus avellana*) has coppicings which are used in the manufacture of hurdles, and sweet chestnut (*Castanea sativa*) is used for making fences.

Coppicing is carried out in winter, so that the whole of the following season is left for new growth, and the cut should be as close to the ground as possible, so that the new shoots grow from the rootstock rather than the cut stem. The cut should slope to one side so that water can run off. The traditional method of cutting was to use an axe or billhook, but a chain saw can be used to speed up the process. Where young trees are planted in order to establish a coppiced area, they should be spaced 2.4–2.75 m (8–9 ft) apart.

4 HORTICULTURAL CROPS

The small, part-time farmer is far more likely to concentrate his crop growing activities on a market garden level than on a farm level. This will enable him to have enough for his own needs, produce marketable crops if he has available local markets and have a surplus for some winter feeding of his livestock.

Equipment

Some form of mechanical cultivator is essential, and these days, there is a wide choice of rotovators, small horticultural tractors and hand tractors with a range of attached implements to cover all the tasks. These include ploughing, harrowing, hoeing and earthing up. Watering equipment is also essential, and this will include one or more standpipes and taps in the garden vicinity. The best type of hosepipe is one which is reinforced to prevent kinking, and which can be conveniently rolled up on a wheeldrum. Those produced for the commercial grower are generally more expensive, but of a better quality than those sold to the amateur gardener. A sprinkler attachment is useful, particularly where seedlings are being established as well as for other plants which require adequate levels of water at particular stages of growth and production. Where crops are grown in glasshouses or polythene tunnels, a trickle system of irrigation is useful. This has a series of outlets which supply water directly to the soil around individual plants such as tomatoes.

A sprayer is a useful piece of equipment for any gardeners, as it can be used to apply non-toxic pesticides as well as chemical sprays. A good handcart is better than a wheelbarrow, as it is lighter to manoeuvre, will hold more and is generally more adaptable. A range of the normal hand tools for the garden is a must, for while some of the original hard work of cultivation can be carried out by a rotovator, there is a wide variety of other tasks which will need the use of a spade, fork, trowel, dibber and hoe.

Some crops such as runner (pole) beans will need supports, while there are all sorts of miscellaneous needs which will become apparent as growing proceeds. These include labels, plant-ties, bird-scaring devices, netting and many other items.

Mechanized cultivation makes ground preparation quicker and easier in a market-garden.

Above: A polythene tunnel greenhouse extends the growing season and makes possible the production of specialized crops.

Below: A variety of early and seasonal crops can be grown in a glasshouse.

Preparing the ground for market garden crops

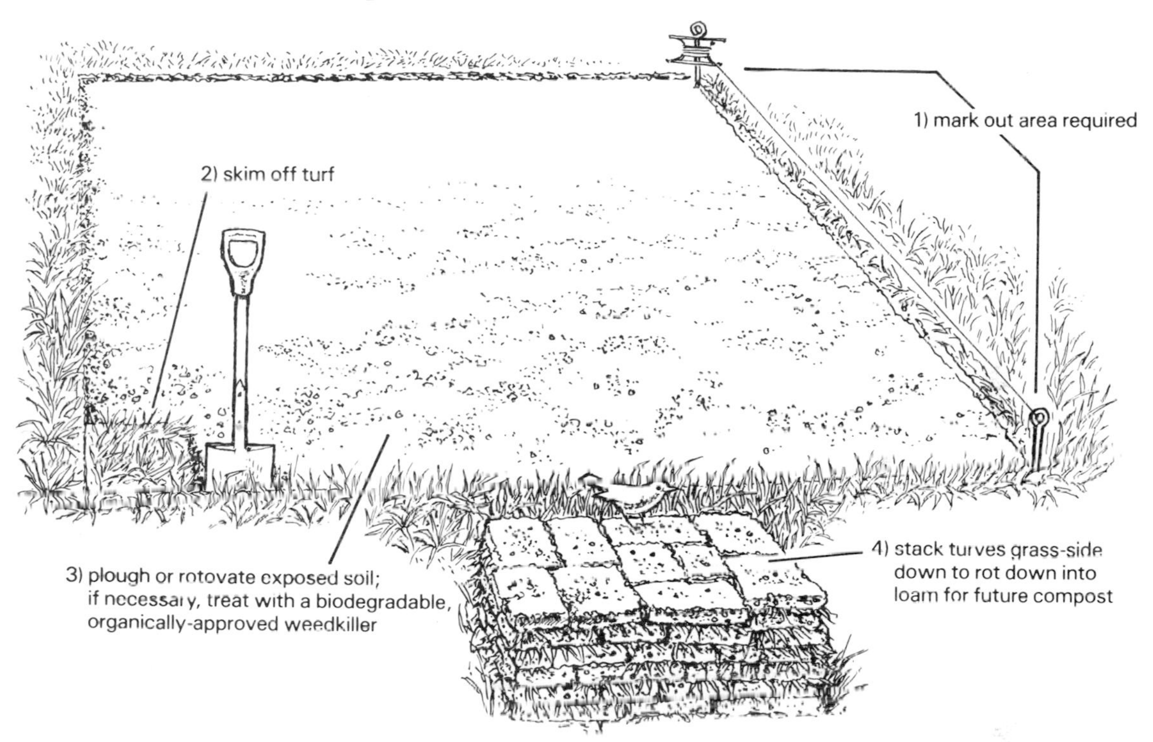

Preparing the ground

If the growing area is being started from scratch it can be ploughed, or it may be a good idea to skim off all the turf from the surface, and stack it grass-side down to rot down into loam soil for making potting compost. Once exposed, the whole area can be rotovated, and well-rotted manure or compost incorporated, to build up the fertility. Where there are perennial weeds such as docks, couch grass, comfrey, marestail and oxalis, it is worth treating the ground with a safe and organically approved weedkiller which is biodegradable. Ammonium sulphamate, which is recommended by the Henry Doubleday Research Association, is efficient against these weeds, as well as a range of annual weeds. It decays in the ground and turns into sulphate of ammonia, adding to the soil's fertility. It is applied at the rate of 450 g to 4.5 litres (1 lb to 1 gal) of water, and this is sufficient to treat 1 sq m (10 sq ft) of ground. It is non-selective, so it is used only for soil clearance, and must not be used amongst plants. Two months must be allowed after application before crops are planted.

What to plant, and where, calls for considerable planning, and it is important to follow a system of crop rotation, so that no crop (apart from obvious perennials such as rhubarb) grows on the same soil two years running. This will lessen the risk of pests and diseases building up in the soil, while allowing nutrients which have been taken out of the ground by one crop to be replaced by another. An example of this is by growing brassicas in a bed previously occupied by legumes. The latter add nitrogen to the soil by the action of nitrogen-fixing bacteria on the root nodules. This nitrogen is gratefully received by the nitrogen-hungry brassicas.

A convenient and effective form of crop rotation is the four-year rotation:

A	B
POTATOES	LEGUMES
C	D
BRASSICAS	ROOTS

Each year there is a changeover: potatoes occupy the place vacated by roots, the legumes take the potato plot, the brassicas move into the space left by legumes, and the roots take over the brassica bed.

MAINTAINING THE SOIL FERTILITY

Where livestock are kept, there is no problem about keeping the soil fertile, but manure must be allowed to rot down before being applied to the soil. The easiest way of doing this is to stack it and cover it with a tarpaulin to avoid all the nutrients leaking out with excessive rain. In a market garden, it is a good idea to compost it with other compostable materials such as grass cuttings. The best way of achieving this is to build proper compost containers which can be used in rotation as rotting down is completed. Regular testing of the soil to see whether there is a lime or other mineral deficiency is to be encouraged, and the appropriate areas of soil treated accordingly.

Constructing a compost heap

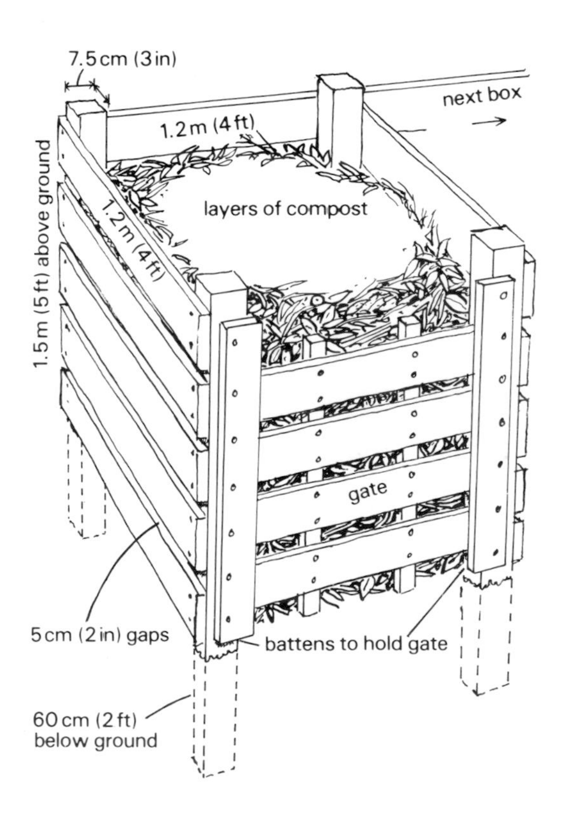

Cash crops

The following is a list of crops which are worth growing for sale, but again, it should be emphasized that only individual market research in your own locality can be the real indicator. It may also be better to concentrate on one or two crops only, and grow any others just for your own use – if you have the time! The list is by no means exhaustive; there may be other crops that some people find it worthwhile to grow commercially.

ASPARAGUS

This is, without doubt, a luxury crop for which there is a consistently high demand during its season, May and June. Traditionally, asparagus was grown on raised beds, incorporating large quantities of horse manure. This method enabled the grower to cut the spear or stem with a long white section of shoot for ease of handling, although most of the white part is tough and inedible and therefore cut off before cooking. The modern method of commercial growing is to use level ground with slight ridges of 5–7.5 cm (2–3 in), but the drainage must be good. Where ground is badly drained, the traditional raised beds are an advantage.

In the autumn the ground is prepared in one of the following ways, depending upon the scale of operations. Spread well-rotted manure on strips 45 cm (1 ft 6 in) wide and 90 cm (3 ft) apart on weeded, cultivated ground, then rotovate it in. This is the method frequently practised in the Evesham area, where there is a long tradition of asparagus-growing. On a smaller scale, dig trenches 45 cm (1 ft 6 in) deep, 45 cm (1 ft 6 in) wide and 90 cm (3 ft) apart, then fill them with well-rotted manure. Farmyard manure is ideal and could come from any of the following sources – horse, cow, goat, rabbit, poultry. The small farmer is usually in the fortunate position of having a plentiful supply of manure. The important thing is to make sure that it is completely rotted before use, otherwise the decomposition of the manure in the soil will not only generate heat, which may damage the roots, but will also lead to nitrogen depletion at the expense of the crop. The best way of ensur-

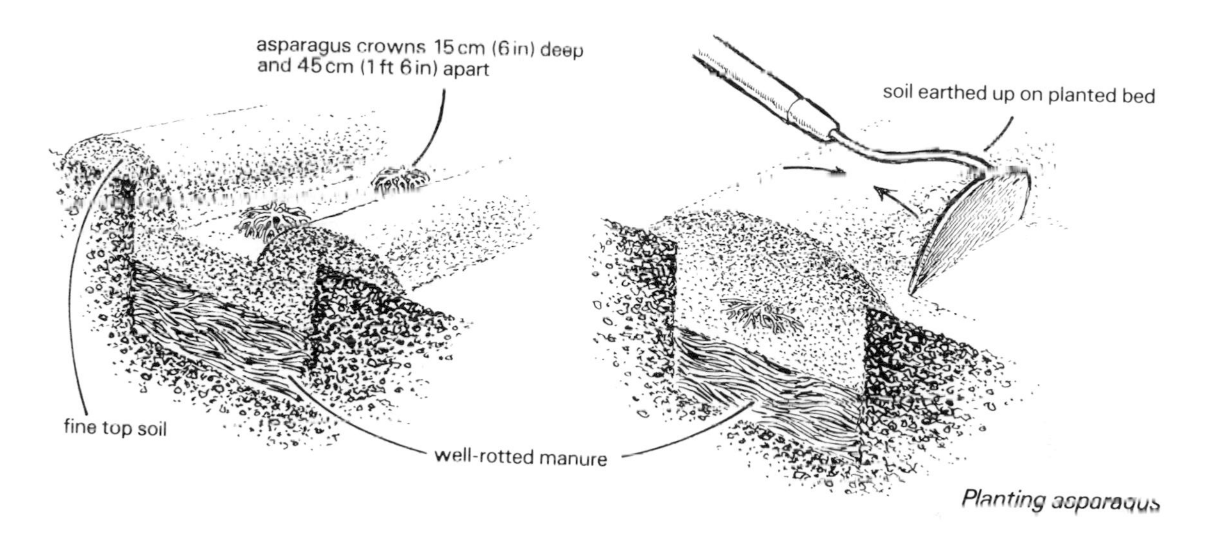

Planting asparagus

ing that the manure is well rotted is to stack it for several months and turn it several times during that period. Add 5 cm (2 in) of good top soil to the manure in the trench then leave it over the winter.

By about February, the level of the manure and soil will have dropped in the trenches and this will need to be topped up before planting, but first, the pH level of the soil should be established. Asparagus dislikes acid soil and is at its best with a pH value of 6 or above. Apply lime at the rate of 55 g per sq m (2 oz per sq yd) for a normal soil, and more for an acid soil, the amount depending on the degree of acidity to be balanced. The rain will then wash it in. Add a little extra top soil to bring the soil level back to what it was, if the level has dropped during the winter, and the beds are now ready for planting.

Two- or even three-year-old asparagus crowns are frequently offered for sale but one-year crowns transplant much more successfully because there is less damage to the fleshy roots than there is with older ones. It is possible to grow asparagus from seed, but this is not worth while for a commercial crop because the wait until a crop is ready is so long: three to four years for asparagus from seed, and two years for a crop from year-old crowns. It is worth bearing in mind that it is the 'strain' of plant that is important, rather than just the variety. Connovers Colossal is the most widely grown variety, and others include Martha Washington and Martha Washington Improved, which are resistant to rust disease. Try to obtain your stock from a reputable asparagus breeder, rather than from someone who merely sells asparagus plants, because in this way you will be assured of good plants bred for reliability and production.

In March or April, depending on weather conditions, fork the top soil in case it has become compacted, then plant the crowns 45 cm (1 ft 6 in) apart, 15 cm (6 in) deep. Make sure that the roots are not damaged and that the crowns themselves are not left exposed before planting. Plant in such a way as to allow the roots to be spread out comfortably. After planting, rake the soil level.

During the first summer, no asparagus shoots should be cut, so that the energies of the plant are concentrated on building up strong plants for future production. Weeds can be kept down by hoeing, and in dry periods watering should be carried out to ensure that the plants grow well without a setback. In the autumn, the feathery fronds of the leaves begin to turn yellow and these should be cut down before the small berries fall, giving rise to unwanted seedlings. Pull out any weeds and fork carefully around the plants. Finally, apply a 5 cm (2 in) layer of rotted manure or compost and leave to overwinter.

In the second year, apply lime at the rate of

55 g per sq m (2 oz per sq yd) as soon as February has arrived, then draw the soil up in a ridge around the plants. As they grow, continue earthing up, either by drawing up the soil with a hoe, or adding new topsoil to the surface. If particularly early crops are required, cloche protection can be used from February onwards.

Cutting can commence at the end of April, depending on the season. The shoots will be fairly small in the first year of cropping and cutting should only last six weeks. After this, the plants should be left to grow on in order to build up stronger plants for the next year. The following season, cutting can last for nine weeks, with the maximum yield being attained after seven years. The bed itself, if carefully tended, should last up to twenty years as a permanent perennial cropping area.

The shoots are cut under the soil, to a depth of about 7.5 cm (3 in), with the ideal length being 25 cm (10 in). During the cutting season, all the growth, even the thin shoots, should be cut. For marketing purposes, the shoots should be graded, with those of uniform thickness being put together, and then bundled in to 450 g (1 lb) weights. Each bundle is tied at the top and bottom and then trimmed level at the base.

Yields will obviously vary, depending upon how long the bed has been established, but for a new crop you should aim to get 450 g (1 lb) of asparagus from every three crowns. For a well-established bed, the yield will approach 450 g (1 lb) per two crowns, although this is presupposing ideal growing conditions and no accidents. The average commercial yield on a field scale is approximately 4 kg (9 lb) from forty-eight crowns.

Asparagus has few enemies (apart from friends who come to stay just when they know it to be in season), and if resistant varieties are grown, there should be no problems with rust disease. Perennial weeds such as nettles and docks will interfere with its growth, and the need for weed-free beds is evident. Asparagus beetle may make an appearance, but it is not a common problem. Where it does appear, it is normally between July and August and a non-toxic and organically acceptable deterrent in the form of derris dust will deal with it.

BRASSICAS

Brassicas undoubtedly do best on relatively heavy soil and there is really not much point in trying to grow them if your soil is light and prone to drought. Cabbages, brussels sprouts and cauliflowers are produced in great abundance by the large growers and I do not feel that there is much scope for the small producer, unless he knows there to be a specific demand in his area, and he is geared to meeting it. Brussels sprouts need a long growing season, while cabbages and cauliflowers are difficult to grow organically because of the large numbers of pests that attack them which are difficult to control without reliance on chemical aids.

There is one brassica, however, which is regarded as an exotic crop and which may be more appropriate to the small specialist grower; this is the kohlrabi, a quick-growing plant that produces edible balls on the stem. It does have some disadvantages - as with all exotic crops, only a minority of consumers will buy it, but this is less relevant to the small producer than it is to the large one. Its advantages are many: it is extremely quick growing and can mature in eight weeks. It can withstand a certain level of drought and is not as liable to clubroot as the other members of the brassica family. The crop must be harvested before it gets too big (no more than tennis-ball size) otherwise it becomes woody and inedible.

The soil should be well cultivated in the autumn, with well-rotted manure or compost incorporated. In spring, lime is added to bring the pH value up to 7.5 for, like all brassicas, kohlrabi likes lime. The seed can be sown in succession from February onwards if the weather is mild, although early crops may be damaged by pigeons and so will require protection. In view of this added expense, it may be more appropriate to concentrate on later sowings only. The seed is sown thinly where cropping is to take place, in rows 30 cm (1 ft) apart, and the seedlings are thinned out to 15 cm (6 in) apart. Weeds should be kept down in between the plants, and an effective way of doing this and conserving moisture at the same time, is to apply a mulch, such as straw, if this is available.

For early crops the varieties White or Green

Vienna are suitable, but for later, winter supplies, Purple Vienna is better.

GARLIC

Garlic likes full sun and a medium to light soil. It also likes a reasonably long growing season, so it is necessary to plant out the cloves in February for there to be a long enough period of growth. So, it is obvious that a cold, northerly area is not suitable for this crop, although it is frost hardy it requires plenty of sun in order to ripen properly. Most of the garlic sold in Britain is imported from Italy and it is certainly bigger than anything that could be grown in this country. It is difficult to compete with Italy when it comes to sunshine, but there is a market for organically produced garlic.

One of the great advantages of growing garlic as a crop is that it has few pests, and indeed, there is a certain body of evidence that shows it has a protective influence in deterring pests from other, nearby crops. The soil should, ideally, be cultivated and manured in the autumn and then limed to bring the pH value to between 7 and 7.5 before planting. The cloves are planted just under the soil surface, about 10 cm (4 in) apart, rows being 25 cm (10 in) apart. Little is required in the way of cultivation during the growing period, apart from hoeing to keep annual weeds at bay, and pinching out any flowerheads which may form. When the growth begins to die down in the summer, the garlic should have the soil scraped away slightly from around the bulbs, so that they can ripen effectively on top of the soil, as is the practice with onions. When the foliage has died down completely, lift the bulbs carefully and leave to dry in a cool, airy place. Once this is done, the bulbs can either be plaited into strings, like onions, or the tops can be trimmed so that they can be sold individually. It is a good idea to remove as much soil as possible from the roots, without damaging them or needlessly removing the white, papery leaf scales.

RUNNER BEANS

The home-produced runner bean has stiff competition from the large producers, and prices will tend to be low if there is a glut during the season. The only way in which competition is possible is by selling yours at a slightly higher price on the basis that they are organically produced. If the scale of operations is large enough, and adequate car parking facilities are available, it may be possible to advertise your runner beans locally on a 'pick-your-own' basis.

Runner beans are a relatively easy crop to grow, as long as the soil is in good heart and has either been manured for a previous crop or during the autumn cultivation. They are susceptible to frost and should not be sown in the open until about the second week of May. Earlier sowings are possible in mid-April under cloches, or in pots in a cool greenhouse for subsequent planting out.

Supports are necessary for the climbing beans (referred to as pole beans in the USA) and these can be arranged in one of two ways: staking in rows or using a wigwam arrangement of poles. For the staking method the seeds may be sown in a double, staggered row, 25 cm (10 in) apart with 15 cm (6 in) between the seeds. The poles are placed firmly in the ground

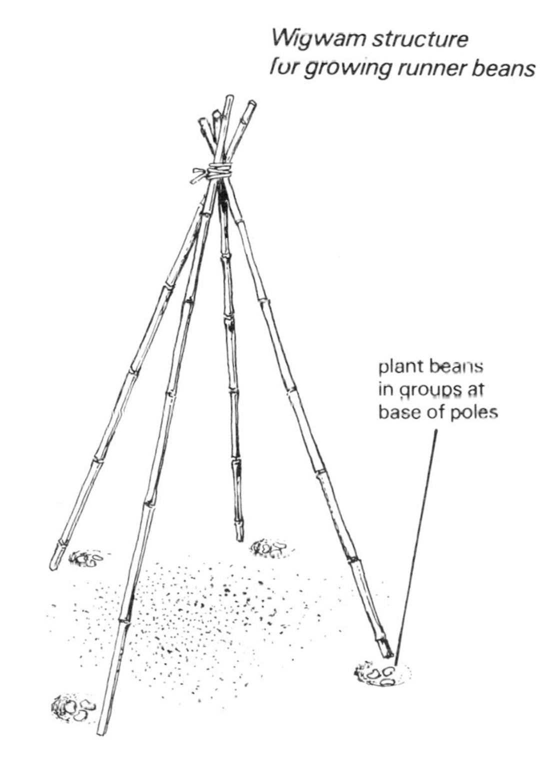

Wigwam structure for growing runner beans

just outside the plants when they come through, then bent inwards and tied to a horizontal pole running across the top. A wigwam structure presents less of a continuous surface area to strong winds, and is made by placing four poles 7.5cm (3 in) apart in a square with the ends tied together at the top. About three seeds or plants are placed at the base of each pole. The disadvantage of this method is that cloches cannot be used early in the season, although, of course, plants can be raised in a greenhouse and then planted out later.

For a commercial venture, the cost of the supports can be high, and it may be that the outlay is too much for a reasonable return. Heavy-grade bamboo poles are satisfactory and will last for several seasons if stored in a dry place through the winter. They are also light and easy to handle.

There are several good varieties available, including Achievement and Prizewinner, which have long pods, and White Emergo, which is particularly useful for freezing and canning. Kelvedon Marvel and Streamline are suitable for growing as ground beans if the tops are pinched out and the plants allowed to branch out.

The plants must be kept well watered, and well composted or manured ground will help to retain moisture. In dry conditions, there is a danger of the flowers falling off before the pods have a chance to form. Slugs may also be a problem, particularly with ground beans, but the main problem is more likely to be aphids. Derris dust should be applied as soon as possible, before the numbers become too great for this to have effect.

FRENCH OR DWARF BEANS

This is another steady seller, particularly if made available slightly early in the season by using crop protection. It adapts well to indoor cultivation such as an unheated greenhouse and will also grow in pots. Cloche protection can also be used. On a field scale, it requires a light, warm soil. If sown too early, when the soil is cold, the seeds will rot, and this is even more likely where there is a combination of a heavy soil and low temperatures. The main enemy is the slug and again, this is often more apparent in wet soils. The dwarf bean can withstand a certain amount of drought, and there is no doubt that dry soils have less slugs. In view of this, it may be appropriate to grow another crop if your soil is a particularly heavy one.

Where cloches are to be used, they should be placed on the ground at least a week before sowing in order to give the soil a chance to warm up. The first sowing can then take place around the middle of April, although this will vary depending upon prevailing weather conditions. The soil will not need manuring, as long as it has been manured for a previous crop, but it should be well cultivated and friable. The seeds should be spaced about 7.5 cm (3 in) apart and 2.5 cm (1 in) deep, in a staggered double row 30 cm (1 ft) wide, but rather than waste surplus plants, they can always be planted out to provide a later crop, or perhaps even sold as seedlings as long as it is possible to get a buyer to collect them the same day as thinning takes place.

Unprotected sowings should not take place until about mid-May, but again it depends on the weather conditions in your particular area. Sowing can either be in staggered double rows as for the cloched crops, or in single rows, which are spaced 60 cm (2 ft) apart. 0.25 litre ($\frac{1}{2}$ pt) of seed will be needed for a 30 m (100 ft) row.

Dwarf beans should not require support, but if winds are frequent, then it may be necessary to stake them with some bushy twigs. These often help to keep the beans off the soil where they might otherwise become soiled, but it should also be remembered that this may make picking more difficult and time-consuming.

Picking should take place before the pods become hard and lumpy. Once harvesting begins, it should continue two or three times a week so that the maximum crop is obtained. If pods are left on, they quickly become stringy and tough, and the overall yield of the crop is reduced. In this respect it may be appropriate to concentrate on stringless varieties rather than on a variety such as The Prince which, while it is a reliable cropper, is extremely stringy. Tendergreen is particularly suitable for forcing under cloches and is good for freezing; it is stringless as long as it is picked young. A good commercial variety is Masterpiece, while an epicure crop is provided by some of the round yellow-podded salad varieties.

Dwarf beans can either be left loose in a carton for weighing out on the spot or weighed and pre-packed in polythene bags with the weight marked on them.

SALAD CROPS

Lettuce are worth growing for sale because they are relatively quick growing, and can be sown in all the odd spare patches of ground or in space vacated by early-harvested crops. If pre-sown in boxes and then transplanted, it is a good idea to cut the end of the root before transplanting. This places less strain on the young plants and enables them to get off to a quicker start. They may need protection against birds, and either cloches or netting can be used. Cloches can also be used to produce early lettuce, which will command a higher price. Crisp varieties such as Webb's Wonderful are popular, as well as Cos types, which stand well and last a long time before going to seed.

Salad onions and radish are also useful for using up spare patches of ground. They are sold in bundles secured by an elastic band. White Lisbon is a reliable variety of salad onion, while the radish French Breakfast is popular. It is also worth trying the variety White Icicle, which is a large, tapering pure white radish that lasts a long time before becoming woody.

TOMATOES

Tomatoes are always popular and there is always a place for the locally produced ones, particularly if they are grown organically. One of the problems encountered by the organic grower is that of combating root diseases such as root rot, nematodes, fusarium and venticillium. The normal grower either uses proprietary 'growing bags' or sterilizes his soil chemically each year, a course of action which is unacceptable to the standards of the organic grower. One way of overcoming this is to graft tomato plants onto resistant root-stocks, which are available from commercial growers. The root-stock is slower to germinate than the fruiting variety and should therefore be sown four days earlier in a temperature of 18°–21°C (65°–70°F). Once growing well, the temperature need only be 13°–15.5°C (55°–60°F). When the plants are 10 cm (4 in) tall, they are ready for grafting. Have the two plants, one on each side of you, and take the root-stock and cut off its top, leaving one or two leaves. Now, using a razor blade, make a downward cut about 12.5 mm ($\frac{1}{2}$ in) long, but not quite through the stem. Take the fruiting variety and, leaving the top on, make a similar cut upwards. Fit the two cuts together so that they interlock, and bind them together with 19 mm ($\frac{3}{4}$ in) wide adhesive tape. It is much easier to do this with a helper, one holding while the other binds. Pot on as one plant, with both root systems intact and leave to grow until large enough to transplant into the permanent position.

The use of cloches is an inexpensive way of producing early crops.

A trickle system of irrigation is best for tomatoes, with water seeping directly where needed around the roots.

If whitefly is a problem, a biological form of control is available that is now used increasingly by commercial growers. This is the mini-wasp *Encarsia formosa*, which is only the size of a pin-head but is parasitic on the young of whitefly. Specialist suppliers make it available as a leaf covered with infected whitefly larvae. This leaf is cut into pieces and placed around the whitefly-infected greenhouse. The chalcid wasps then hatch out and become predators of the whitefly, effectively containing the numbers, but some sticky deposit will form on most fruits. A wipe with a damp cloth will remove this after picking.

Red spider mite, which can also be a major pest in protected conditions, is similarly controlled using a predatory mite, *Phytoseiulus persimilis*. This is normally supplied on bean leaves, which are distributed around the greenhouse at the first sign of infestation.

Tomatoes are picked before they are quite ripe, so that handling does not cause bruising, and they last longer once harvested. They are normally weighed out and sold in bags, although, if being supplied through a retailer, flat cardboard containers are frequently used.

MUSHROOMS

Any empty outbuilding can be adapted and used as a mushroom house, and it is a project worth investigating if you have a ready source of horse manure. The manure needs to be stacked and allowed to rot before being used to make up the beds. Gypsum at the rate of 12.7 kg (28 lb) per ton of manure is mixed in to counteract the acidity. The beds are made up in the outbuilding, as long beds down either side, with timber walls 25 cm (10 in) deep to contain them. An alternative is to use boxes, 15–20 cm (6–8 in) deep and stack. This is economical in space and makes for ease of handling. The manure is placed in the beds or boxes and well firmed. When the temperature of the manure is at 21°C (70°F) it is time to introduce the 'spawn'. A soil thermometer placed deep into the manure will indicate the temperature.

Commercial suppliers sell cultures of sterilized mushroom spawn and this is introduced as pieces about the size of a walnut 2.5 cm (1 in) deep in the manure, at 25 cm (10 in) intervals. Firm the manure and ensure that it is damp, without being too wet. In about a week, long white mycelium threads will be seen growing on the surface of the manure. Put a 3.8 cm ($1\frac{1}{2}$ in) deep layer of a good, sterilized potting compost on the manure; a peat-based one is ideal. Make sure that the compost is slightly dampened and, as long as the temperature is round about 13°C (55°F) and does not fall below 7°C (45°F), mushrooms should begin to appear in approximately six weeks. If the tem-

Boxes for growing mushrooms stacked on greenhouse-type staging

perature is likely to fall drastically, a 15 cm (6 in) layer of straw can be used to give protection to the beds.

The mushrooms are picked regularly and sold either in punnets or in bags by weight. After cropping is complete, the compost can be used to build up the fertility in the vegetable garden.

SOFT FRUIT

Raspberries, strawberries, gooseberries and blackcurrants are all popular with customers. Some part-time farmers and growers confine their commercial crops exclusively to soft fruit production.

Raspberries do best in a moist, cool environment. They need well-cultivated soil with plenty of organic manure incorporated, and good drainage. It is important to plant certified disease-free canes, and the MAFF publishes a list of certified growers for this purpose. Glen Cova and Malling Jewel are good varieties.

The canes will need supports, and a convenient way of arranging this is to have posts and wire, with straining posts to give added firmness at the ends of the rows. In dry periods, overhead spray irrigation will be needed to ensure proper formation of fruit.

The fruit is picked regularly, and it is important to handle it with care in order to avoid damage. Raspberries are normally sold in paper punnets. Once the fruit is all picked, the fruited canes are cut back to the ground, leaving unfruited, green canes to grow and mature for the next season.

Strawberries are an important crop, particularly for the 'pick-your-own' market. It is important to obtain clean, healthy and virus-free stock and to plant in deeply tilled and manured soil, with good drainage. Reliable varieties are Cambridge Favourite and Royal Sovereign. Young maiden plants obtained from a reputable supplier are planted in late summer or early autumn. It is important to plant properly, ensuring that the roots are adequately spaced out so that they can grow without restriction. Early crops are possible by using polythene tunnel cloches, but it is important to watch out for slugs in this protected environment.

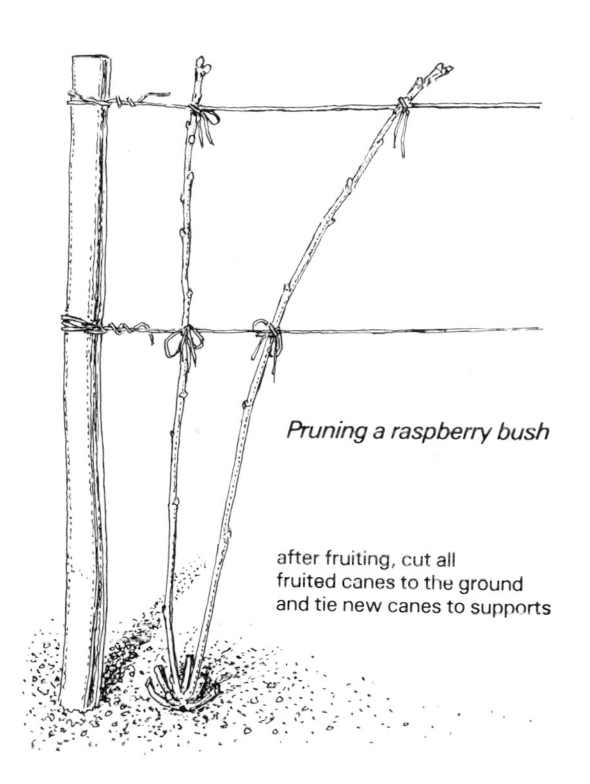

Pruning a raspberry bush

As the plants grow, soil is hoed up around them because they make new roots higher up the crown. This can be done by hand, or mechanically if the strawberries are grown on a field scale.

When the plants flower, straw is placed along the rows so that the fruit will be protected from the soil and by mud splashed up by the rain. All runners should be removed. The fruit is picked as soon as it is ripe, and is normally sold in punnets. After fruiting is complete, the straw is often burnt where it stands so that insect pests, weeds and diseases are destroyed, and a small amount of potash is made available for the soil. The strawberry plants themselves are not damaged by this treatment and come up again as if nothing untoward had happened. A strawberry bed will last for three to four seasons. After that it is best to make a new one.

Gooseberries are normally grown on a 'leg', which allows cultivation under the plants to take place, and which facilitates the removal of unwanted 'suckers'. Two- or three-year-old bushes are planted in the autumn at a distance

Pruning a gooseberry bush

before pruning an established bush – most fruit will be carried on new wood, so cut as little new wood as possible

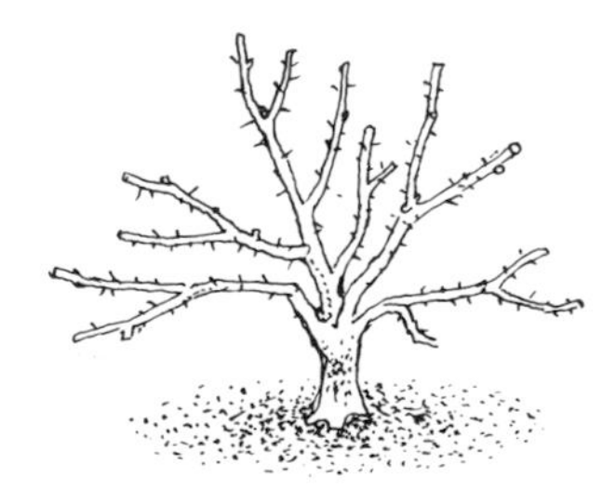

after pruning – tangled and overcrossing shoots have been removed to keep bush open – lateral growth on old wood has been cut back

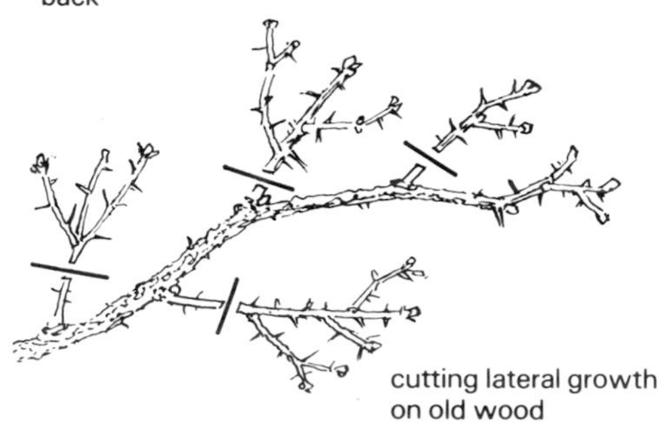

cutting lateral growth on old wood

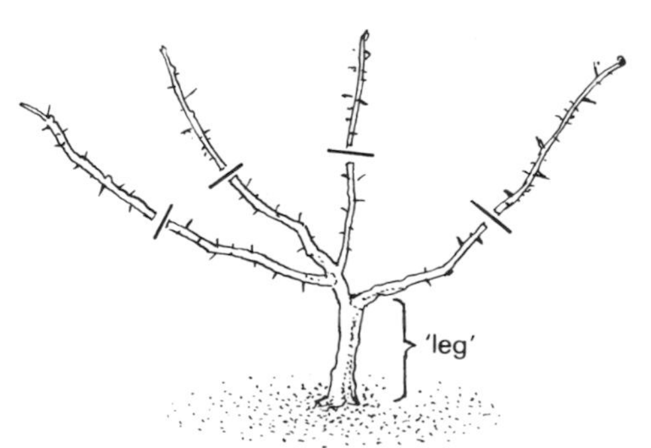

to prune a newly-planted bush, cut main shoots to half-size to encourage good growth and subsequent shape of bush

of 1.5 m (5 ft) each way to allow for subsequent growth and spread. In the first year, the leaders or growing stems will need to be cut back by half so as to form a strong framework for future growth. Once they are established, the bushes will only require regulatory pruning, which means keeping the centre of the bush open and removing crowded or rubbing branches.

Blackcurrants are planted as one- or two-year-old bushes, 1.5 m (5 ft) apart each way or in rows, 1.8 m (6 ft) apart, with 90 cm (3 ft) between the rows. The latter makes for easier access for pruning and harvesting. They are somewhat susceptible to frosts and are best protected by hedges or windbreaks. Immediately after planting, the branches should be cut down to ground level to ensure strong new growth in the spring. In subsequent years, pruning should be to remove about one-third of the branches so that the bush is kept open and vigorous.

Gooseberries and blackcurrants are permanent bushes that do well on rich, well-cultivated soils with plenty of organic manure. In fact, it

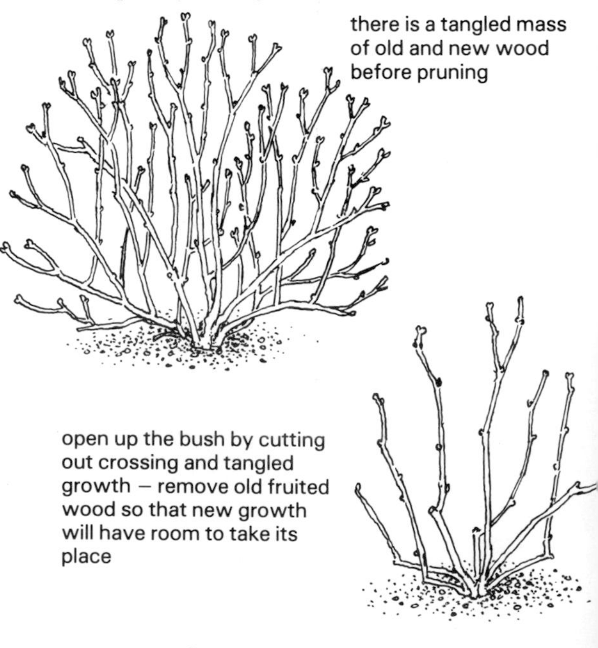

there is a tangled mass of old and new wood before pruning

open up the bush by cutting out crossing and tangled growth – remove old fruited wood so that new growth will have room to take its place

to prune a newly-planted bush, cut back hard to encourage plenty of new growth

Pruning a blackcurrant bush

seems to be the case that a good crop of blackcurrants is impossible without organic manure. Good varieties of gooseberry are Leveller and Careless. Seabrook's Black, Wellington XXX, Boskoops Grant and Baldwin are reliable varieties of blackcurrant. Both gooseberries and blackcurrants are sold in punnets.

TOP FRUIT

Dessert and cooking apples and pears, plums and damsons can all find a ready market. Establishing fruit trees for commercial purposes, however, is costly and there is a considerable period of waiting before a harvestable crop is available if new trees are planted. Anyone thinking of planting an orchard for commercial purposes should take the advice of the divisional MAFF office and also enquire about the possibility of a horticultural grant.

Cider manufacturers will often buy a whole season's crop of cider apples and perry pears and it is worth investigating this as a possibility if you have such trees. If you have five acres or more given over to apples and pears which are not for cider and perry making, and you intend selling the fruit, it is necessary to register with the Apple and Pear Development Council.

VINEYARDS

Vineyards for the commercial production of wine grapes are becoming increasingly popular in Britain, with the southern counties providing the necessary degree of sunlight. Vines grow well on sloping, stony and poor ground, so it is possible to utilize land that is no good for anything else. The early varieties Muller-Thurgau and Reichensteiner grow best in Britain, for the later-maturing one cannot be guaranteed to mature in the relatively short summer.

Grafted and healthy plants are planted 1.5 m (5 ft) apart, in rows 90 cm (3ft) apart along a south-facing slope so that they receive the maximum amount of sun. Where machinery is used, the rows may need to be wider apart to allow access. Posts with supporting straining posts are placed at the end of the rows, with two strands of wire running between them, the first at a height of 45 cm (1 ft 6 in) and the second at a height of 30 cm (1 ft). Planting takes place between November and March. The vines are pruned back to three buds immediately after planting. As growth proceeds, it is tied to the supports, and grapes are produced on the previous year's growth, so no crop can be expected in the first year after planting. Even then, it is best to cut the fruiting bunches off so that the energies of the plant are concentrated on strong root formation. A moderate crop can be taken from the third year after planting.

A convenient way of training the vines is to use the 'double arch guyot' method. This involves tying two fruiting canes along the bottom wire, in each direction. The next winter, after harvesting, these canes are cut back and are

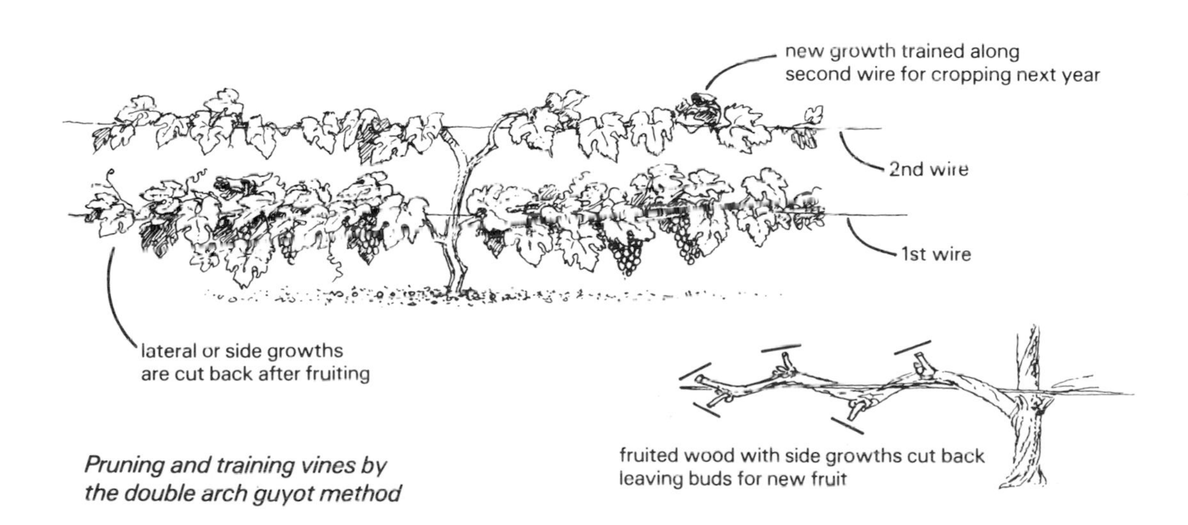

Pruning and training vines by the double arch guyot method

replaced by two new ones. The fruiting canes are always trained along the bottom wire, while the new growth is trained along the top strand.

If you wish to make your own wine for your own private consumption from your grapes, you can do so without any restriction in Britain, but if it is to be sold, then a certificate from Her Majesty's Customs and Excise is required. Those interested in investigating the possibilities of vineyards should contact The English Vineyards Association.

Vines need careful pruning and training on supports.

Cut flowers

The cut flower trade has become highly specialized, with customers either buying direct from a shop, or arranging to have flowers sent from a distance through the auspices of organizations such as Interflora. The sign 'Cut flowers for sale' outside a private house has become a rare sight, but this does not indicate that there is not a ready market for local buyers. In the USA, it is not unknown for self-sufficient communities to rely on the sales of cut flowers and garlic as their main income source. However, they operate on a fairly large scale and fulfil trade contracts, rather than sell in small amounts to the general public.

On a small scale, the easiest approach is to concentrate on selling bunches of mixed flowers, rather than taking orders for particular flowers. This means that you can make up the bunches from whatever happens to be available at a particular time, and if one crop fails it is not a disaster.

The important things to bear in mind are to ensure a steady supply of flowers for as long a season as possible, and to grow foliage plants as well. Attractive foliage enhances and provides a suitable backdrop to the flowers. On a more economical note, it also means that you can cut down the number of flowers in the bunch. Long-stemmed varieties are easier to cut and are generally more popular with customers.

PREPARATION OF THE SOIL

The plants will be either perennials, where seasonal growth continues from the same roots year after year, biennials, which grow in the first year and flower in the second, or annuals, which need to be sown every year and die down at the end of the season. The perennials may be herbaceous, the top growth dying down in the winter, leaving only a root or bulb, or shrubby which, as the term implies, are shrub plants. Many of the foliage plants are shrubs.

The ground where *perennials* are to be grown should be completely cleared of weeds, particularly the troublesome couch grass. If this is left, it will compete with your plants for nutrients and will adversely affect their root growth. Where ground is badly affected, it should be treated with a weedkiller such as ammonium sulphamate, which is biodegradable. Leave for two months after application then rotovate the ground and incorporate well-rotted manure. Let the ground settle for a week or two and then plant the perennials, leaving enough room between them to allow for subsequent spread. Apply lime if the pH value indicates an acidic imbalance. Where rotted manure is not available, bonemeal is a good and long-lasting substitute.

Biennials, such as wallflowers and sweet william, are best sown in well-cultivated ground in rows, and thinned out as the seedlings grow. Again, the ground should be weed free, otherwise perennial weeds will make life a nightmare. The flowers will appear in the second year, so if replacement plants are needed, they should be sown every year, otherwise you will only have these flowers once every two years.

Annuals are hardy or half hardy. The former can be sown outside in March or April, depending upon prevailing weather conditions, but the latter cannot be sown until mid-May, or planted out until June, when the risk of frost has gone.

PLANT SUPPORTS

Growing long-stemmed flowers for ease of cutting and to keep the customers satisfied is one thing, but it does mean that the plants are at the mercy of the winds, and will sprawl over unless adequately supported. Providing supports can be costly, and this is a factor which needs to be taken into consideration in overall costings.The plants do best in full sun, but if hedges or fencing provide windbreaks, without casting shadows, this is the ideal situation.

Stout posts hammered into the ground at each end of the row will enable wire or horticultural twine to be stretched along to provide support, but larger plants may require individual support in the form of bamboo canes. There are a variety of supporting aids available and these are generally advertised in the gardening press. Sweet peas will grow up a structure such as is used for runner beans or they can be trained up wigwam poles.

CUTTING

If the plants are grown in convenient rows, walking between them to cut the flowers is simplified. As a general rule, the more frequently cutting takes place the better, for if seed pods are allowed to form, the number of flowers is reduced. It may, however, be more economic of time to set aside a cutting time once a week. Small secateurs are suitable for cutting, but make sure that each flower is cut with as long a stem as possible. The old fashioned trug is very good for carrying the flowers while you cut. It allows each flower to be placed horizontally, without crushing other plants. Keep each variety separate and place in water in a cool, dark outhouse. The bunches can then be made up the day before selling. Small elastic bands or raffia can be used, but take care that the bunches are not tied too tightly otherwise the stems may be crushed. Thin paper is usually used for wrapping up the bundles and special paper for the horticultural trade can be purchased if the level of sales warrants it. As for what price to charge, only you can be the judge of that. Check what local shops are charging and then relate this to your own expenses and time.

RELIABLE PLANTS FOR CUT FLOWERS

The list of possible plants and varieties is endless, so I have restricted it to include those plants which seem to be best for the purpose: relatively long-lasting once cut, and popular with customers. Individual varieties are not listed, for there are too many and new ones are being introduced each year. It is a good idea to study the seed catalogues thoroughly before making a choice. Finally, gardeners in northern areas may find some of the half-hardy annuals difficult to grow, particularly in wet, cold summers, and this is another factor to be borne in mind.

Foliage plants

Plants which are grown predominantly for their leaves and which provide a useful addition to bunches of mixed flowers:

Senecio greyii – greyish foliage

Spirea 'Anthony Waterer' - hardy perennial; useful spring foliage
Cotton Lavender (*Santolina chamaecyparissus*) - grey leaves.
Cineraria maritima - silver foliage.
Lambs' ears (*Stachys lanata*) - silver-grey leaves.
Cypress (*Cupressus lawsonia*) - dark green, evergreen foliage.
Golden-leaved Alder (*Alnus incana aurea*) - golden, variegated leaves.

Perennial plants
Achillea filipendula Gold Plate'
Alkanet (*Anchusa italica*)
Chrysanthemums
Golden Marguerite (*Anthemis tinctoria*)
Shasta Daisy (*Chrysanthemum maximum*)
Michaelmas Daisy (*Aster Cultirus*)
Coreopsis lanceolata
Dahlias
Larkspur (*Delphinium elatum*)
Garden Pinks and Carnations
Leopards Bane (*Doronicum plantagineum*)
Purple Cone Flower (*Echinacea purpurea*)
Avens (*Geum coccineum*)
Gladioli
Helenium autumnale
Iris sibirica
Liatris spicata
Veronica spicata
Phlox paniculata
Rudbeckia spp.
Rose (*Rosa floribunda* spp.)
Scabious (*Scabious caucasia*)
Purple Loosestrife (*Lythrum salicaria*)
Bergamot (*Monarda didyma*)
Lupin (*Lupinus polyphyllus*)
Lychnis chalcedonica
Golden Rod (*Solidago canadensis*)
Montbretia (*Tritoma* spp.)
Daffodils and Narcisshus
Tulips

Biennials
Wallflowers (*Cheiranthus cheiri*)
Sweet William (*Dianthus barbatus*)
Stocks (*Matthiola incana*)
Canterbury Bells (*Campanula medium* - Cup and Saucer varieties)

Annuals
Snapdragon (*Antirrhinum major*)
African Daisy (*Arctotis stoechadifolia grandis*)
Pot Marigold (*Calendula officinalis*)
Cornflower (*Centaurea cyanus*)
Clarkia elegans
Cosmos bipinnatus
Star of the Veldt (*Dimorphotheca aurantiaca*)
Godetia amoena schaminii
Sweet Pea (*Lathyrus odorata*)

CATERING FOR FLOWER ARRANGERS

The art of flower arranging is increasingly popular and the provision of dried flowers, leaves and grasses for this purpose could bring in an income for a green-fingered entrepreneur. The material can be sold direct to interested individuals or through craft equipment suppliers and craft shops. It is worth contacting the local Women's Institutes and local authorities who may be running courses in various crafts. Advertisements in local newspapers are often useful.

Sprays of beech leaves, particularly copper beech, are always in demand. Cut the twigs just when the leaves are at their best, and before they begin to decline in the autumn, and place immediately in a solution of 1 part glycerine (anti-freeze) to 2 parts water. Leave them to stand in this way, in a cool place until the solution is drawn up the stems and into the leaves: the presence on the leaves of an oily deposit indicates when this has been achieved. Remove the twigs from water, dry the ends and gently wipe the leaves to bring up the sheen.

Everlasting flowers (*Helichrysum* varieties) are perhaps the simplest ones to grow for flower arrangers. They really only require sowing and cutting because their papery petals are already so dry that they resemble living dried flowers. Either sow them in a greenhouse and transplant out of doors in May, or sow direct outside in May to June. Gather the flowers when they have reached a good size, but before the petals have opened sufficiently for the yellow centre to be seen. Put the stalks together in uneven bundles and hang upside down in a dry, airy place where there is no danger of damp. Remove brush foliage from stems before drying.

Sometimes only the flowers are required, either for making dried flower pictures or for attaching to artificial wire stems which allow them to be placed in any position. In this case, the stems can be removed and the flower heads placed on wire mesh or stretched muslin trays where the air can circulate to them. Do not, however, do this unless you know you have a buyer, for most people will require their flowers with the stalks still attached.

The species of perennial statice (*Limonium incana* and *Limonium latifolium*) are useful. The former is white and can be dyed, while the latter is purple, and although it does have a tendency to lose some of its colour during drying, nevertheless it is suitable for a wide range of displays.

Teasel (*Dipsacus sylvestris*) is particularly popular with flower arrangers specializing in Christmas decorations. The spiky heads can be easily sprayed with varnish or gold paint, and last for a long time. It grows readily from seed,which is available from many seed companies, and, as long as it has a reasonable level of moisture in the soil, will grow without help from anyone. In fact, once you have it in the garden, it will self-seed and reappear every spring. Initially, it can be sown either in boxes and transplanted outside as seedlings, or direct into the ground during March or April. Cutting the stalks and spiky heads takes place in autumn and all that is necessary is to to remove any obviously wrinkled or disfigured leaves. Teasel is usually treated as a biennial.

Herbs

Herbs are popular with many people and they are relatively easy to grow. As most of them are hardy perennials, there is no need for much in the way of artificial heating.

One of the most effective ways of selling herbs, or indeed any plants, is to have a garden where prospective customers can see them growing. The soil does not need to be particularly rich, otherwise the plants grow too lush and lose much of their aromatic quality, but it does need to be well drained. It must be said, however, that the number of plants sold in this way will be limited to the number of callers. Most herb suppliers operate a mail order business, and some even specialize in dealing exclusively with the trade. On a small scale, a combination of all these is possible. Callers can be catered for, local garden centres, horticultural suppliers, and even greengrocers can be approached to see if they will sell your herbs. At first, they may agree to do so on a sale or return basis only, but if there is a demand, this could change.

One of the problems in selling herbs is that customers often do not know what to do with them. Despite the popularity of books on herbs, there is generally a lack of information. Gone are the days when housewives relied on their herb patches for 'simples' or home cures for family ills, and natural flavourings and additives for their cooking. These days, most people are familiar with mint and parsley, and that is it. It is almost essential, therefore, to provide a descriptive label with each herb sold, giving basic details of planting, cultivation and uses. If the number of plants sold is small, these could probably be handwritten or even duplicated on a simple copier, but for large numbers, it is obviously better to have them printed professionally. Printing is extremely expensive, and it is a good idea to get quotations from as many local printers as you can find. There are also some who operate on a mail order basis and these usually advertise in publications such as *Exchange and Mart*.

Plant pots and potting compost will be necessary. The former can be rigid plastic or flexible black plastic, and the latter must be sterilized to kill off any weed seeds and soil pathogens. If you cannot produce your own potting compost then it will need to be bought. The difficulty with home-produced compost is efficient sterilization of the loam soil. If this can be overcome, an adequate formula is as follows: 2 parts sterilized loam soil, 1 part sand, 1 part peat for seeds; 7 parts sterilized loam soil, 3 parts sand, 2 parts peat for potting. Soil sterilizing equipment is used by many nurseries, and if the level of business warrants it, it may be appropriate to buy it. Horticultural material of this kind is regularly advertised in the horticultural trade press.

Some of the more popular herbs

A Annual B Biennial P Perennial

Name	Propagation	Height	Uses
Angelica (B)	Sow in July to flower following year.	1.5–1.8 m (5–6 ft)	Stems cut and candied for confectionery.
Balm (P)	Sow in spring. Divide roots in winter.	90 cm (3 ft)	Infuse leaves for tea, and use in salads.
Basil (P) (best treated as half-hardy annual)	Sow in pots in spring. Does best inside in sunny position.	30 cm (1 ft)	Chop leaves and sprinkle on tomatoes.
Bergamot (P)	Divide roots.	90 cm (3 ft)	Infuse leaves for tea.
Borage (A)	Sow in August.	90 cm (3 ft)	Blue flowers used in salads or candied for confectionery
Burnet (P)	Sow in spring. Root division in autumn.	60 cm (2 ft)	Use leaves in salads.
Camomile (P)	Sow in April.	30 cm (1 ft)	Make tea from infused flowers.
Chervil (A)	Sow in April.	60 cm (2 ft)	Aniseed-flavoured seeds used in soups.
Chives (P)	Sow in spring. Divide bulb cluster in winter.	23 cm (9 in)	Use as a garnish, and in salads.
Coriander (A)	Sow in April.	90 cm (3 ft)	Seeds used in soups, pickles and curries.
Dill (A)	Sow in April.	90 cm (3 ft)	Leaves used in salads and in sauces to go with fish. Seeds used in pickles and vinegars.
Fennel (P)	Sow in spring.	1.8 m (6 ft)	Leaves used for sauces with fish. Swollen stem cooked as vegetable.
Hyssop (P)	Sow in spring. Cuttings in late summer.	60 cm (2 ft)	Good bee plant; can be clipped as hedge. Leaves used in pot pourri.
Parsley (P) (treated as annual)	Sow in April, then every six months, to have plants in sequence.	30 cm (1 ft)	Garnish in salads, sauces. Will grow in shade.
Lavender (P)	Sow in spring. Cuttings in autumn.	35 cm–90 cm (1 ft 3 in–3 ft) depending on variety	Clipped as hedge. Flowers and leaves in pot pourri and lavender bags.
Marjoram (P) (annual variety also available)	Sow in spring. Divide roots in autumn and also take cuttings.	30 cm (1 ft)	Use leaves in soups, salads and stuffings.
Mint (P)	Sow in spring. Take root cuttings in autumn and pot up.	60 cm (2 ft)	Used for mint sauce, and as a garnish.
Rosemary (P)	Rooted cuttings.	1.2 m (4 ft)	Leaves sprinkled on meat before roasting. Also used for hair rinses.
Sage (P)	Sow in spring. Take cuttings in late summer.	60 cm (2 ft)	Good for bees, and in stuffings. Useful as gargle for sore throats.
Summer Savory (A) (Perennial Winter Savory is a different plant)	Sow in April. Take cuttings of winter variety in autumn.	30 cm (1 ft)	Use leaves to flavour meat and sausages. Use leaves of winter variety as pepper substitute.
Sweet Cicely (P)	Sow in spring. Divide roots in winter.	90 cm (3 ft)	Useful for stewed fruits and tarts, allowing sugar in recipe to be reduced. Will grow in shade.
Tansy (P)	Sow in spring. Divide roots in winter.	90 cm (3 ft)	Leaves used very sparingly for cakes and stuffings. Good shade plant.
Thyme (P)	Sow in spring. Cuttings and root division.	7.5–23 cm (3–9 in) depending on variety	Leaves used with meat and in soups and stuffings.

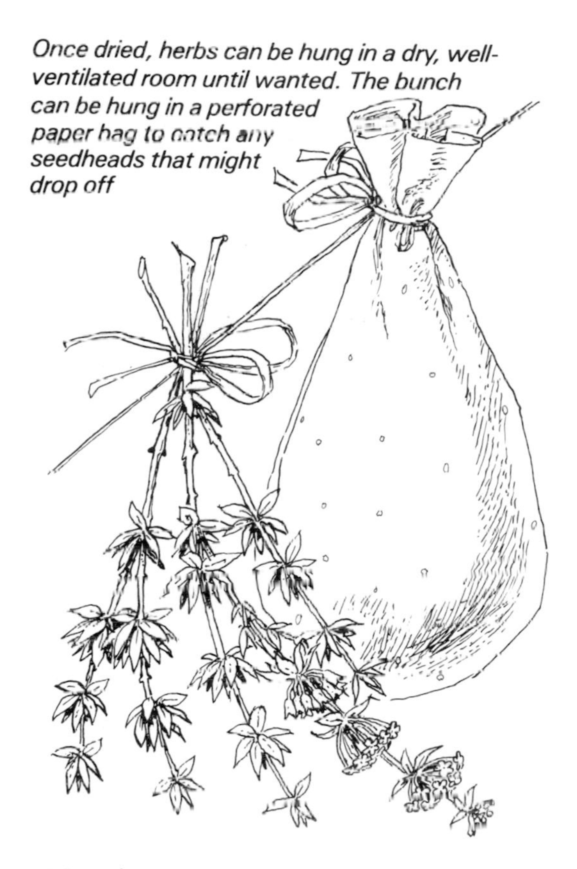

Once dried, herbs can be hung in a dry, well-ventilated room until wanted. The bunch can be hung in a perforated paper bag to catch any seedheads that might drop off

Most herbs can be sown from seed in the spring, while clumps of perennials can be divided in the autumn. Cuttings of perennials are also easy to take. It is simply a matter of taking shots with a 'heel' from the main stem, and putting them in a sandy compost. Hard wood cuttings may get off to a better start if the ends are dipped in water first, then in hormone rooting powder before being placed in the compost. The cuttings are left in a cool, moist situation, out of direct sunlight, and when rooted, are transferred to individual pots.

Herbs can also be sold dried and there is a wide range of herbal products that can be produced for sale. These include pot pourri, lavender bags, herbal soaps and candles.

The British Herb Society is a useful organization for those interested in any aspect of herbs, while in the United States, the organization is The Herb Society of America.

Other plants

A glance at the classified advertising columns of the gardening press indicates that a great variety of plants are sold by mail order or to callers. These include hardy perennials, house plants, alpines, bulbs, miniature roses and many others. The cost of mail order is now much higher than it used to be and most businesses are specialist ones.

In the spring and early summer, there is a demand for trays of bedding plants such as African marigolds, petunias, lobelias, alyssum and salvias. Seedling vegetables such as lettuce, leeks, cabbage and tomatoes are also bought in great numbers, and this could be a useful source of subsidiary income for the gardener who has greenhouses and is geared to this type of production. Whatever area of plant growing is envisaged, it is a good policy to go and visit existing nurseries and enterprises, to find out how they operate. The Horticultural Trades Association is a useful organization for those involved in this field.

5 THE DAIRY COW

The keeping of any livestock brings responsibilities and restrictions on one's time, but dairy animals are an even greater tie. They need to be milked twice a day and there is a necessary emphasis on clean, hygienic conditions for all stages of milk production. No-one should attempt to start a commercial enterprise involving dairy animals, without first keeping one or two animals for home production, and gaining experience of their needs. Demand, distribution and finance should also be looked at closely.

It is debatable whether a commercial herd of dairy cows can be regarded as a part-time occupation, and it is a question which can only be resolved by the individual after he has examined the situation in detail. It is certainly true that, in parts of the country, notably in the west where small herds of about twenty cows are still kept, an increasingly important source of income is coming from the tourist trade, and the income from the cows is regarded as a subsidiary. Even for a small herd such as this, however, the costs in terms of equipment such as a pipeline milking unit, and the specialized buildings required, are very high. Availability of good pasture is another crucial factor.

The regulations in relation to selling cow's milk and milk products are stringent, and sales are not allowed, even on a tiny scale, unless the dairy farm is registered for this purpose. It is only after the buildings have been inspected and the water tested, and found to be acceptable to the standards of the Milk and Dairies (General) Regulations 1959, that registration will be approved. A copy of the regulations is available from the local MAFF office or from Her Majesty's Stationery Office. The requirements include a regularly cleaned milking parlour with a separate milk room to which the milk is transferred. This milk room must be free from contact with any place where animals are housed, and must be well lit and ventilated. The floor is required to be impervious, while floor, walls and roof should be dust-proofed. The approach to the milk room, where milk collection vehicles arrive, must be a concrete one, and it must be kept clean. The water supply is tested for purity and all equipment used must be regularly cleaned, using approved cleaning agents. After registration is approved, inspections and testing of milk samples are still carried out at regular intervals.

The first step, therefore, is to apply to the local MAFF office for registration as a dairy farm. Once inspection has taken place and registration has been approved, it is necessary to contact the Milk Marketing Board at Thames Ditton in Surrey. This body is responsible for the sales of milk by producers in England and Wales. If the milk is to be sold by wholesale it must be sold to the Board, who determines its destination and arranges for its collection by bulk milk tanker. If a producer wishes to sell his milk by retail, direct to consumers or caterers, he must apply to the Board for a licence to do so – if his sales exceed 225 litres (50 gal) a year.

For the small farmer, a dairy enterprise does at least offer the security of a guaranteed monthly cheque for the milk produced. The question is whether it is sufficient recompense for the time and work involved.

Breeds

The Friesian is the most widely kept dairy cow in Britain, mainly because the volume of milk produced is high, although the butterfat content is low by comparison with that of the Channel Island breeds, the Jersey and Guernsey.

Above: Friesian cows with a Hereford bull; the male progeny will make good beef calves.

Below: The Jersey is the most popular house-cow with part-time farmers.

The Dairy Shorthorn is suitable as a milk-cow as well as a producer of beef although it has declined in numbers in recent years. The Ayrshire, originally bred in Scotland, also produces rich milk.

For a relatively docile cow to produce milk in quantity, and have good beef qualities, the best breed would be the Friesian. If, however, the emphasis is to be on milk products such as cream, yoghurt or soft cheese, then either the Jersey or Guernsey would be the best choice. The snag with the Channel Island breeds is that their milk yield is smaller and they produce bull calves which have little commercial value by comparison with those of other breeds. The beef aspect of dairying is important, for a substantial proportion of all our beef supplies come from fattened bull calves, surplus heifers or culled dairy cows. With the ready availability of artificial insemination, dairy heifers can be crossed with a beef breed bull to produce excellent beef animals. The following are suitable crosses for beef:

Aberdeen Angus × Friesian
Hereford × Friesian
Hereford × Dairy Shorthorn
Beef Shorthorn × Ayrshire
Charolais × Ayrshire

Some breeders have also crossed Charolais bulls with Jersey or Guernsey cows with satisfactory results, but this is still fairly uncommon.

Brucellosis and tuberculosis

Brucellosis and tuberculosis are both serious conditions which can affect cattle and man. At one time they were far more common in British herds, but, thanks to a rigorous testing and eradication scheme, are becoming rarer. Tuberculosis testing is carried out regularly, particularly in areas such as the south-west of England where it is still endemic. One of the points of conflict between the MAFF and ecologists is the claim that badgers are carriers of bovine tuberculosis. Where cattle are found to have tuberculosis, they are compulsorily slaughtered, but compensation is available.

Brucellosis is gradually becoming eradicated on a county-by-county basis. Where an area is free of the disease, all cattle are regularly tested. In areas where it is not yet eradicated, herds can be tested and if, after a period of time, they are found to be free, they are designated as an 'accredited herd'. There must be no contact with non-accredited stock or they will lose this designation. It will be seen, therefore, how important it is to regulate the movement of livestock. Every farmer and livestock keeper is required to keep a record of all movements of stock to and from his premises. In certain instances it is also necessary to obtain a 'movement permit' from the local MAFF office.

Buildings

In the not-too-distant past, it was common in winter for cows to be housed overnight and milked in the same building. There are still small herds where traditional buildings have been adapted, but the modern method is to have a milking parlour separate from the overnight shelter. The milking parlour may be a static one, or rotary where there is a circular revolving platform to facilitate the movement of the cows in and out of the parlour. Within these types, there are a number of variations, depending on the layout of cubicles. Abreast is where the cows stand side by side as in a traditional cow shed; they are either on the same level as the operator or raised above him by about 45 cm (1 ft 6 in). Tandem is where cows stand head to tail about 75 cm (2 ft 6 in) above the operator's pit. Herringbone is where the cows stand at an angle, about 75 cm (2 ft 6 in) above the operator's pit, so that the diagonal lines resemble a herringbone pattern. If a new dairy farmer is considering the installation of a milking parlour, he will glean much valuable help, advice and information from the dairy adviser of the local MAFF. It is also worth attending one of the specialist dairy shows organized under the auspices of the MAFF. Here, the whole range of types will be on show, and demonstrations will be available.

For the small, part-time farmer, the cost of new buildings will probably be so high as to

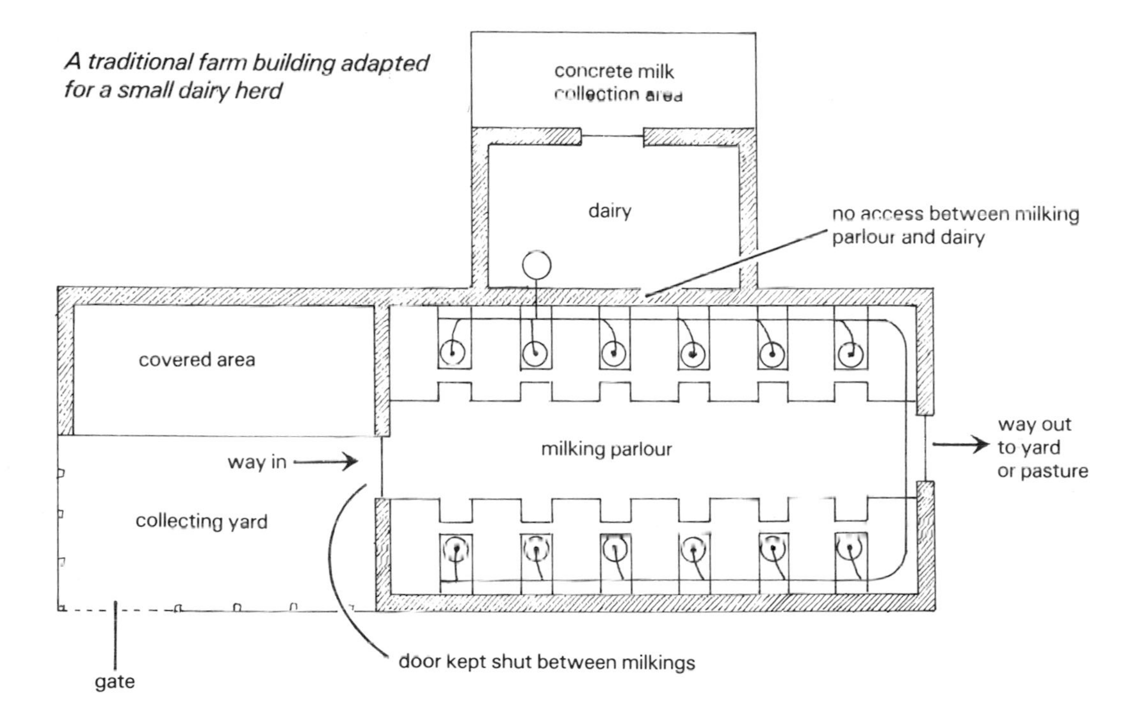

A traditional farm building adapted for a small dairy herd

doom the enterprise to failure from the beginning. There are, of course, grants available, and the local MAFF will advise on these. Perhaps the best course of action for the small farmer would be to adapt an existing cowshed as a milking parlour, and to provide overnight and winter housing in a partially covered collecting yard. This is a fenced-off area with one section roofed for weather protection, and with straw on the floor. Where a yard is used, the regulations require that there be a door separating the yard from the milking parlour, unless the same degree of cleanliness can be ensured on both sides of the opening to the parlour. The door should be kept closed between milkings. The cows are then bought in from pasture to the yard, but if it is winter, they will already be in the yard. They are allowed in through the door into the parlour where they stand side by side on a concrete ramp about 45 cm (1 ft 6 in) above the ground facing inwards to where feed is available. The traditional buildings normally have chain yokes to secure them. Once milking has finished, the cows go back to the yard, either to stay there overnight through the winter, or to go through the yard to their grazing pasture. The milk is transferred to the dairy for filtering and cooling. To comply with regulations, this should have a separate access where no animals are allowed.

Milking equipment

Hand milking was a regular feature of dairy farms when I was a child, but now no commercial enterprise could consider anything other than a machine-milking system. Again, there is a wide variety and the different manufacturers will provide information and demonstrations. The principle of machine milking is that a vacuum created by a pump is produced in the cups enclosing the cows' teats and this is similar to the sucking action of a calf. The level of pressure producing the vacuum should be 3.4 kg to 2.5 sq cm ($7\frac{1}{2}$ lb per sq in) which is equivalent to 37.5 cm (1 ft 3 in) mercury. Regular checking is necessary to ensure that this is not exceeded, otherwise the cups can be pulled up too high, causing damage to the cows' teats. Milk is extracted from the udder, and is then transferred

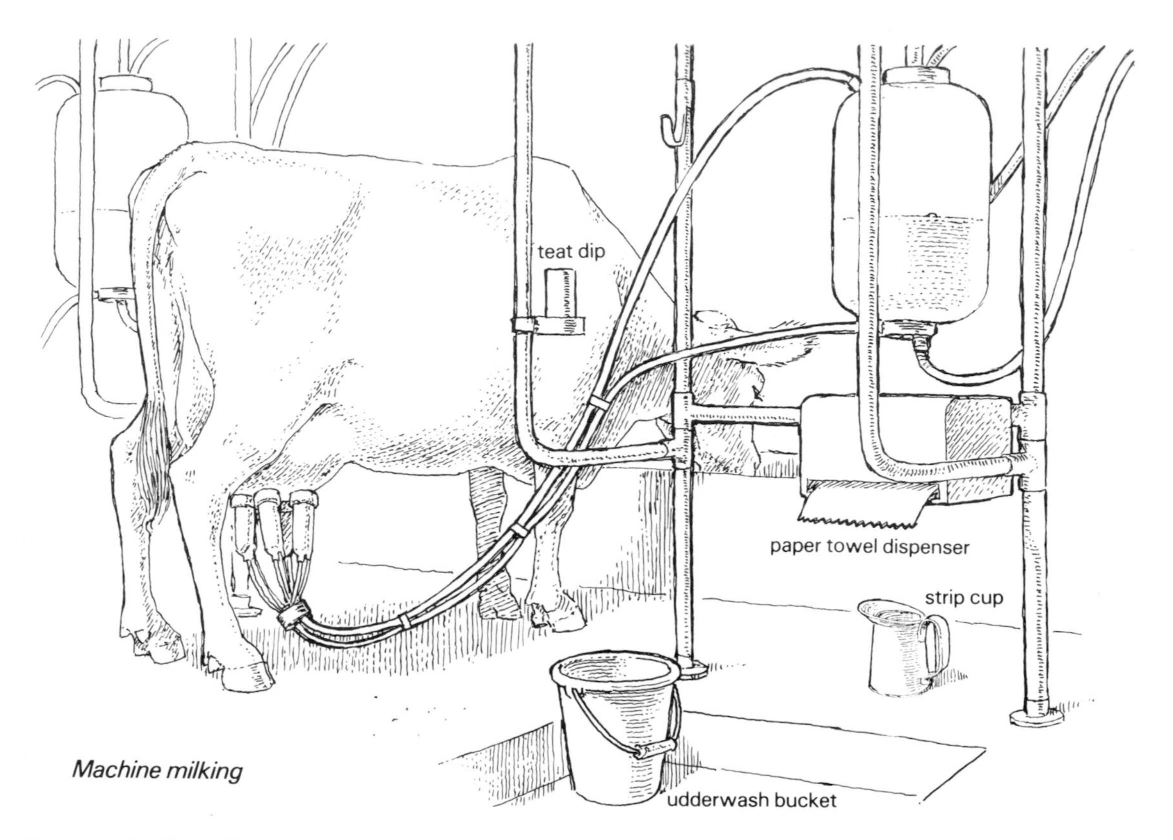

Machine milking

along a pipeline, first to graduated recording jars and then to a central refrigerated tank or vat to await collection by the bulk milk tanker. The recording jars are to allow measurement inspection and the taking of samples. The cows are brought in to the milking parlour and fed concentrates according to yield. The first necessity is to ensure that the udder is clean and free from mud or other contaminants. Udder cloths and buckets of warm water are used, although the ideal, from the hygienic point of view, is a warm water spray. Disposable paper towels are used to dry the udder, and for this as well as other operations, the handler should be equipped with rubber gloves. An overall, rubber boots and head covering is also required to ensure that hygienic conditions are maintained.

After washing, the individual teats are squeezed to extract the foremilk, which may have a high percentage of bacteria. It also ensures that the teat opening is not blocked. The foremilk is squeezed into a special examination cup so that the milker can examine it for any signs of blood or clots which may indicate mastitis, a bacterial infection of the udder.

After the foremilk examination, the teat clusters of the milking machine system are attached to the teats and milking commences until the milk flow ceases. The teat clusters are then removed and individual teats are dipped in a container of dilute hypochlorite or similar antiseptic, to give protection against mastitis.

All the equipment must be cleaned using approved cleaning agents. These are laid down in the Milk and Dairies (General) Regulations 1959, previously referred to. In addition, the MAFF divisional offices have a booklet entitled 'A Guide to Clean Milk Production', which is well worth reading.

The milk, once transferred to the milk vat in the dairy is filtered and cooled, and kept at a temperature of 4°C (39°F) until the bulk tanker collects it each day.

Feeding

It is obvious that no cow will produce milk unless she is adequately fed. The most important source of food is grass, and this is why most of the dairy herds are in the west, with its abundant rainfall, producing lush pasture. Grass is only available from the spring to late summer. Once it has stopped growing, and winter has arrived, the grass must be fed in a dried or preserved form, either as hay or as silage. Fodder crops such as kale, cabbage, mangolds, swedes, sugar beet tops and pulp are grown for cattle feeding. Cereal and pulse crops are made available, as well as mixed corn, and mixtures of peas, beans, oats and barley in varying proportions. These days, a proprietary dairy ration is available for in-milk animals; this contains all the necessary minerals. Water is essential at all times, and a cow in full milk production, can drink up to 68 litres (15 gal) a day.

A cow needs a 'maintenance' ration, which is an adequate supply of food to keep her in good condition when she is not in milk. Once she starts her lactation she will need a 'production' ration, which is the 'maintenance' ration, plus an extra amount, depending upon her level of production. In commercial herds, this is estimated as accurately as possible in order to be as cost effective as possible in relation to feed and production costs. Individual cows do, of course, have different requirements and a large Friesian will consume more than a small Jersey. The genetic make-up of the animal is also important, which is why it is essential to acquire cows from a good milking strain.

The amount of food that any cow will eat is limited by its digestive capacity. If it ate only bulky food such as hay it would be replete once it had reached its capacity, but its nutritional needs in relation to milk production might still be unsatisfied. For this reason, it is important to control the amount of bulk given, and to ensure that concentrates are given in the right balance. This balance will depend upon the level of production. The higher the volume of milk produced, the greater the amount of concentrates, in relation to bulk foods. As an example of this, a large cow such as a Friesian, with an average weight of 558.8 kg (11 cwt) would require the daily winter feeding rates given below.

Hay and silage can vary immensely in nutritive value. A commercial farmer would have his hay and silage analysed, either by ADAS or by firms selling dairy food. The ration would then be formulated for each main stage of lactation.

To a novice, 9 kg (20 lb) of hay may be quite meaningless because he cannot judge by eye how much that is. If this is the case, there is only one way to learn and that is to weigh it out so that the amount of hay involved can be seen. The same applies to silage, or indeed to any feeds. Once the eye has seen the bulk in relation to weight of a particular feedstuff, the owner can then remember it, and hopefully be in a position to estimate by sight in future. A convenient way of weighing for this purpose is to use a spring-balance with a section of canvas or old sheet tied at the corners and suspended from it. Another way of estimating is to bear in mind that one bale of hay weighs 20 kg (45 lb).

A smaller cow, such as a Jersey, with an average weight of 406.8 kg (8 cwt) would require the daily winter feeding rates given below.

Daily winter feeding rates for a Friesian

daily yield	grass silage	hay	concentrates
9 litres (2 gal)	22.6 kg (50 lb)	9 kg (20 lb)	
13.5 litres (3 gal)	9 kg (20 lb)	10 kg (24 lb)	3.1 kg (7 lb)
22.5 litres (5 gal)	9 kg (20 lb)	7.7 kg (17 lb)	6.3 kg (14 lb)
32 litres (7 gal)	9 kg (20 lb)	4.5 kg (10 lb)	9.5 kg (21 lb)

Daily winter feeding rates for a Jersey

daily yield	grass silage	hay	concentrates
9 litres (2 gal)	9 kg (20 lb)	7.2 kg (16 lb)	2.2 kg (5 lb)
18 litres (4 gal)	9 kg (20 lb)	2.7 kg (6 lb)	6.8 kg (15 lb)
22.5 litres (5 gal)	4.5 kg (10 lb)	1.3 kg (3 lb)	9 kg (20 lb)

In summer, when good, fresh grass is available for grazing, silage will not be needed, although concentrates and a small amount of hay will be necessary to balance the laxative effect of young grass. The amount of concentrates will again depend upon the level of milk production, and are as indicated previously. Where grass is restricted in the summer, it should be balanced by the feeding of hay. It is important to feed a proportion of hay in the mornings before the cows are let out to pasture in the early spring. At this time, the new grass has its most laxative effect, leading almost certainly to scouring or diarrhoea in the cows. Barley straw can be used instead of hay for this purpose, as well as for supplementing the grass later in the season. The grass intake itself can be restricted by the use of electric fencing.

CONCENTRATES

The concentrates referred to are normally mixtures of high-energy grains or grain-based compound feeds. The latter are available from feed suppliers as proprietary dairy cattle rations. They contain balanced nutrients, and minerals, including the important mineral magnesium, which is often in short supply in new spring grass. A deficiency of this can lead to a condition known as hypomagnesaemia or 'grass staggers'. Mineral licks also supply magnesium in addition to a range of other minerals. Where concentrate mixtures are made up on the farm, a suitable mix would be as follows:

1 part oats
2 parts barley or flaked maize
1 part kibbled beans or soya bean meal

To provide the necessary mineral supplements, the following is added to every 1 cwt of the concentrate mixture:

450 g (1 lb) common salt
450 g (1 lb) ground limestone
225 g ($\frac{1}{2}$ lb) sterilized (steamed) feeding bone flour

SILAGE

Silage is fermented grass and is one of the most common winter feeds. The most convenient way of making it available to cattle in the winter is to control their intake by the use of an electric fence. If you have previously weighed out a given weight of silage in order to see how much bulk this represents, you will be in a position to estimate how much to allow for your small herd of dairy cows per day, and place the electric fence accordingly. Details of silage-making are given on page 40.

FODDER CROPS

Fodder crops are crops, other than hay and silage, which are grown specifically for the feeding of livestock. One of the most important of these is kale, which can be grown on a field scale, and then made available during the winter by using an electric fence to control the grazing. Other examples are field beans, rape, cow cabbage, lucerne, and cereal and legume mixtures. Also important are the root crops, turnips, mangolds, swedes and fodder beet.

These crops are suitable for a wide range of livestock, as well as cattle, and details are given in the table on page 46–7. It is important to remember that for many animals, root crops must be cut up before they can be eaten. Goats, for example, cannot cope with whole mangolds. Crops such as dried field beans also need to be ground or kibbled, and for small amounts, a hand mill is ideal.

Breeding

For the small farmer, the keeping of a bull is not a practicable proposition. Bulls can be dangerous, need special penning and expert handling, and are unlikely to pay for their keep in a small enterprise. With the availability of artificial insemination, there is no need to keep a bull. All that is necessary is to contact the local MAFF to obtain the address and telephone number of the nearest AI Centre.

Before the AI man comes to do the service, he will obviously need to know when the cow is 'on heat', a condition known as 'bulling'. This is usually indicated by fretful restless behaviour, an almost continuous mooing and either trying to mount other animals or standing still when other cows try to mount her. This period is not without its dangers. My husband was

A normal birth with the calf emerging head and front legs first.

once bending over, clearing out a water tank in a field when a bulling Jersey heifer landed full square on his shoulders! Fortunately he was able to escape her amorous attentions without harm. Another sign of 'bulling' is a slight colourless discharge from the vulva.

At the first signs of 'heat' it is vital to contact the AI centre without delay.Once the vulval discharge has become slightly blood-stained, usually after one or two days, it is already too late and you will have to wait until the next heat about twenty-one days later. The cow should be tied up so that when the AI man comes he will be able to administer the frozen semen as quickly as possible. Usually there is no problem, but with some animals it may be necessary for a helper to hold up the tail. Once the semen is placed in the vagina, the AI man will write out a certificate, and for this he will need to know the cow's name and ear number, which is the number allocated to the farmer's herd. If the insemination does not take, it is normal to allow a second servicing free of charge. If, however, there is a failure to conceive after this, it is wise to call in the vet in case there is something wrong with the cow.

Pregnancy lasts for about nine and a half months, although there may be a week or so either way. During this time the cow needs her normal maintenance ration, together with a ration to cater for her current level of milk production. This will gradually taper off, until about two months before the calf is born, she should be dried off. This must be done carefully to avoid mastitis. The principle is to milk less frequently, as the less that is drawn out, the less is produced.

About four weeks before calving, the udder will begin to fill out, and later signs immediately before the birth are a slackening of the muscles on either side of the tail, while the vulva itself enlarges. Calving can take place outside, particularly if the weather is fine, and there is usually less danger of infection in these conditions. If the cow is brought inside, it should be put in a stall that has been well scrubbed out and supplied with clean, fresh straw. There are usually no problems with calving, but if there are, the vet should should be called immediately.

For the first few days, it is essential for the calf to suckle its mother. In this way it is able to have the colostrum or first milk which is particularly rich in nutrients, and also carries many anti-bodies from the mother. It has a good start in life and is protected against infection until it is able to produce its own anti-bodies.

From the fifth day onwards, the calf must be separated from the mother and put in its own pen. There is no reason why several calves should not be penned together although some farmers do not like doing this. Experience shows, however, that they respond far better to separation from the mother if they have other company.

The calf will need to have a milk substitute twice a day up to the age of six weeks, and this needs to be made up according to the manufacturers' instructions. It will also need to be taught to drink rather than suck. The way to do this is to have a bucket of milk ready, then give the calf your finger to suck. Gradually lower your hand until it is in the bucket and as soon as the calf's nose reaches the milk it takes milk in as well as your finger. Gently remove your hand once it has started to drink by itself. It will snort and sneeze for a while, while it gets used to this new way of drinking, but is usually too greedy to stop drinking in order to find your finger again. Hay, fresh clean water and calf concentrates should be available from one week onwards.

From the age of six weeks onwards the calf can be weaned from milk and fed hay, fresh clean water and a proprietary calf weaner ration. Where the latter is not given, a mixture can be made up as follows: 5 parts flaked maize, 3 parts crushed oats, 1 part soya meal, 1 part milk powder. During the next few weeks, the amounts are increased from 85–115 g (3–4 oz) a day to about 1.3 kg (3 lb) by the age of nine weeks.

When the weather is mild the calves can be allowed out to graze on young, fresh grass, but the grazing must be controlled, and should not exceed a couple of hours for the first week or two. This is to avoid the condition known as scouring, which is brought on by an excess of lush grass in the spring.

Opposite: After being thoroughly licked and cleaned by its mother, the calf will soon be on its feet.

The period for which calves are kept will depend on the individual farmer. Some will sell them at the age of a few weeks, while others will be kept and raised for beef or as replacements for the older cows. Whatever the fate of the calf, however, it will need to be castrated if it is a bull calf. This is best done by a vet, unless an experienced livestock handler is available.

Vaccination against a number of diseases is possible, but it is worth talking to the vet and getting his advice as to what protections are necessary for your particular area. If the calf is a female, and it is intended to breed from her, she should be vaccinated. Under the brucellosis eradication scheme in Britain, this is available free of charge, and is normally carried out at about the age of three months. Calves which are vaccinated in this way will have an identifying metal ear tag. Identification is important and can be by either ear tags or ear tattooing.

Ear tagging is an effective means of identifying livestock.

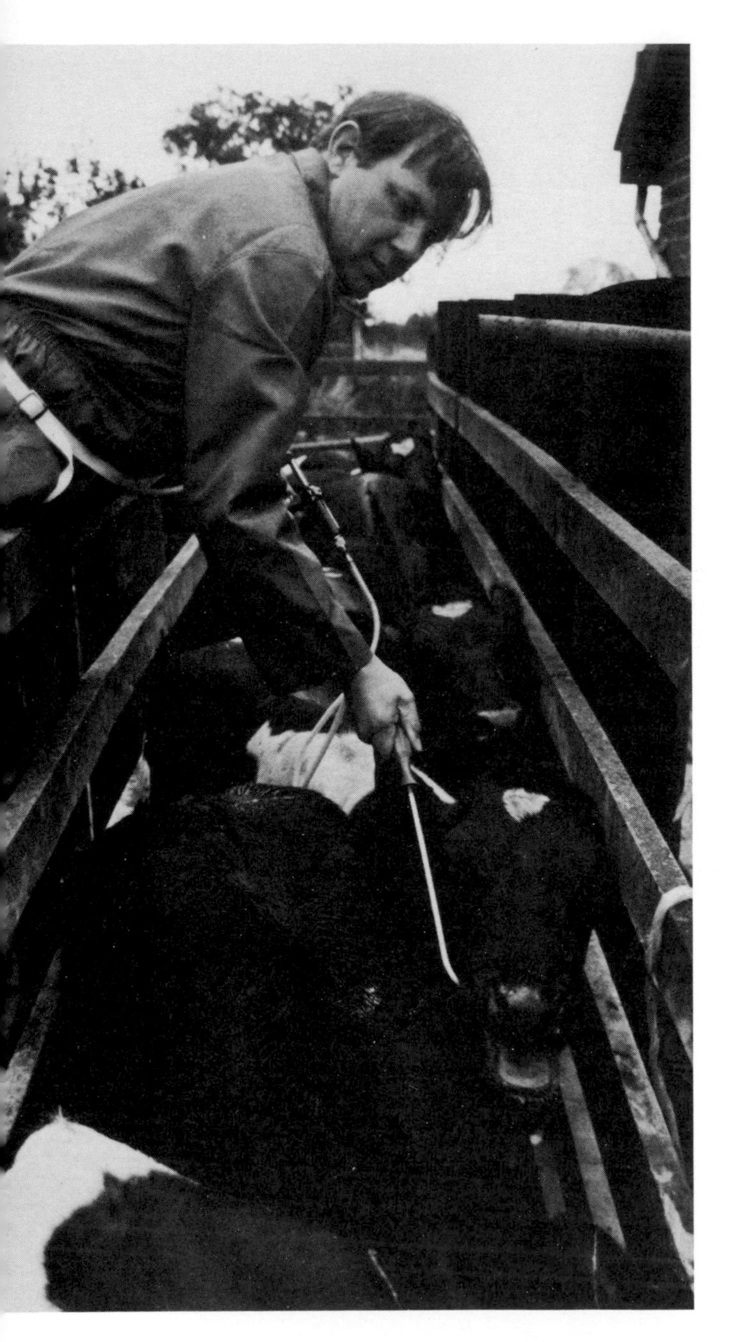

A drenching gun being used to administer a vermifuge for the control of parasitic worms. This is simplified if the livestock are properly confined.

Disbudding is advisable, but should only be carried out by an expert, and with the use of local anaesthetic. The latter is required under the Protection of Animals (Anaesthetics) Act 1954, in Britain. The best way of carrying out disbudding is to use an electrically powered dehorning iron which burns out the buds on the head.

Health

Reference to brucellosis and tuberculosis has already been made earlier in the chapter. There are certain notifiable diseases in relation to cattle, and these include cattle plague, anthrax, foot and mouth disease, pleuro-pneumonia, and rabies. The latter is not in Britain at present, thanks to a rigid policy of animal control in and out of the country.

Any sudden and unexplained illness, particularly where more than one animal is affected, should be reported to the vet immediately. He will also advise on what vaccination may be deemed appropriate in your particular area.

Worming should take place regularly, as well as treatment of the skin with insecticide to deal with lice, mites and warble fly. Scouring or diarrhoea is usually the result of too much lush grass. It is best treated by giving warm water to drink and hay only to eat for a day. If it does not improve in twenty-four hours, the vet should be called in case an antibiotic is needed.

Reference has already been made to the importance of preventing mastitis. When it does occur, it will need antibiotic infusions administered into the teat aperture of the affected quarter. It is not difficult to do, but should be demonstrated by a vet or other experienced person first. In a dairy animal the milk should continue to be drawn from the affected quarter and discarded. No milk should be sold from the animal until the condition has cleared up and three clear days have elapsed since the last antibiotic injection.

Which dairy animal?

Animal	Advantages	Disadvantages
Cow	Hardy in winter conditions. Easy to confine. Artificial insemination readily available. High milk yield, so labour costs per litre lower. Cow's milk readily accepted by public. Calves have commercial value. Guaranteed price for sale of milk. Much advice and information available.	Expensive to buy. Expensive buildings and equipment required. Shorter lactation period (nine months). Stringent regulations apply to commercial milk production. More land needed than for other dairy animals. Good quality pasture needed. Good quality hay needed. Subject to more diseases than goats.
Goat	Relatively cheap to buy. Simple housing and equipment is adequate. Long lactation period (up to two years). Surplus milk can be frozen. No stringent regulations apply to sale of goat's milk and milk products. Less land needed. Browser not grazer, so poorer quality grazing is adequate. Poorer quality hay acceptable. Not subject to brucellosis or tuberculosis. Milk products can attract premium prices in minority whole-food market.	Can be difficult to confine and may need tethering. Destructive to hedges and trees. Not hardy in wet or winter conditions. No artificial insemination available yet. Lower milk yield than cow, so labour costs per litre higher. General public has negative attitude towards goat's milk. Male kids have little commercial value. Very little commercial advice or information available.
Milk sheep	Milk high in butterfat. Good for making quality yoghurt and cheese. Products can attract premium prices in minority wholefood and delicatessen market. Provide wool 'crop' as well as milk. When crossed with breeds such as Southdown, give good meat lambs which have commercial value.	Not readily available. Relatively expensive to buy. Not winter-hardy Delicate feet prone to foot rot. More prone to mastitis. Artificial insemination available only as a limited service. Use of ram is more common. More difficult to milk by hand (but can be machine-milked). General public do not know about sheep's milk. Stringent health regulations require regular dipping. Good-quality pasture needed. Good-quality hay needed. Very little commercial advice or information available.

6 GOATS & MILK SHEEP

The dairy goat

The goat is remarkably free of official attention in Britain. Unlike in France and Switzerland, where it has always played a major role, the goat has been largely ignored by British farmers, and indeed is still regarded by many as an object of mirth. This attitude, coupled with the widely held public belief that goat's milk tastes 'funny', has led to a situation where goatkeeping is largely the prerogative of an informed minority. Only in recent years has the increasing interest in health and wholefoods focused attention on goat's milk as a suitable alternative for those who suffer from allergies. For young babies, it is also reputed to be better because its composition is more like that of human milk than cow's milk.

Official overlooking of the goat in Britain happens not only because it is a minority livestock, but also because, unlike the cow, it is not subject to tuberculosis or brucellosis. The result is that, apart from the basic health regulations which apply to all foodstuffs, there are no specific dairying regulations which apply to goat's milk and its sale. This makes it ideal for the small, part-time farmer, who wishes to utilize his surplus milk in a profitable way.

BREEDS

There are several breeds of goats available, including the Saanen, British Saanen, Toggenburg, British Toggenburg, British Alpine, Anglo-Nubian, British and Golden Guernsey. The best breeds from a dairying point of view are those which have been developed from good milking stock. In Switzerland, the Saanen and Toggenburg are the main dairying breeds, but in this country their yields are not generally as high, mainly because of past shortage of breed-

A British Alpine is a good breed for milk yields.

A Breed Champion Saanen Milker.

Toggenburg.

The Anglo-Nubian, whose milk has a higher butterfat content than that of other breeds.

ing stock. The British Saanen and the British Toggenburg are bigger and, in many ways, improved versions of the originals, and these together with the British Alpine, would be the best breeds for milk yields. Of course, it is not quite as simple as that, as it is equally important to buy goats from stock that is known to have a good milking record. For this reason, it is better to buy British Goat Society registered stock from a known breeder who will be able to supply details of parentage. Where goat's milk with a higher than average butterfat content is required, the Anglo-Nubian breed is a better choice, although the overall yield of milk may be lower.

Female kids find a ready market, particularly if they have been de-horned after birth, are registered with the British Goat Society and have been earmarked for identification. Male kids generally have little commercial value, although they can be raised for meat and slaughtered at the age of five months. There are indications, in Britain, that there is a growing demand for goat's meat from the Saudi-Arabian population in and around London and also from Indian restaurants, and it could be that this particular avenue is worthy of exploration.

The interior of a well laid-out goathouse, showing individual penning.

BUILDINGS

Goat housing needs to provide dry, sheltered conditions from the weather, and provided these are met, it does not need to be a complicated or expensive building. Ideally, each milking goat should have its own pen equipped with hay-rack, mineral lick and water bucket. It is not difficult to partition an existing building in this way, and an individual goat soon learns which is her pen. The floor can be kept covered with straw bedding, with clean straw added as necessary, and mucking out need only take place once every few weeks, when there is a build-up. This manure produces excellent compost when rotted down in a compost heap. Milking should take place in a separate area, and a platform on which the goat can stand will make hand-milking easier. A platform is not difficult to construct for anyone with a reasonable knowledge of carpentry, or purpose-built ones are now available. It is usual to feed concentrates to the goats while they are being milked, so that they are happily occupied.

MILKING

Milking can be done by hand (which is common in the case of goats), or by machine if the number of animals warrants the expense. There are several companies who sell milking equipment specifically for goats, and the principle is exactly the same as that utilized for cows. With hand milking there is an extra responsibility to ensure clean, hygienic conditions, and the milker should wash his hands and wear a protective overall. Long hair should be tied back. The udder is wiped and the foremilk taken and examined for any evidence of mastitis, then milking proceeds as quickly as possible. For hand-milking there is really no alternative to good quality stainless steel buckets. They are certainly more expensive than plastic or aluminium ones, but they do not become scratched like plastic, and last a lifetime. They are also easier to clean. Immediately milking is finished, the milk is filtered and then poured into a storage container. A purpose-made filter unit is available for hand-milkers and this is used in conjunction with disposable paper filters. The ideal container into which to pour the milk during the filtering process is the traditional churn, but since the change-over to bulk tanker collection, these churns are no longer being manufactured. There are suitable alternatives, however, such as metal containers manufactured for the food and drinks industry.

The next step is to cool the milk, and the easiest way of doing this is to place the container in a deep sink and run cold water over it, unless you have a traditional in-churn cooler. Once cooled, the milk is ready for transferring into waxed paper containers or purpose-made plastic bags, if the milk is to be sold, and then stored in a refrigerator until distributed. Milk for freezing should be clearly marked with the date. If the milk is not being sold, but is to be used for yoghurt or soft cheese production, it can be left in its metal storage container in a refrigerator until needed.

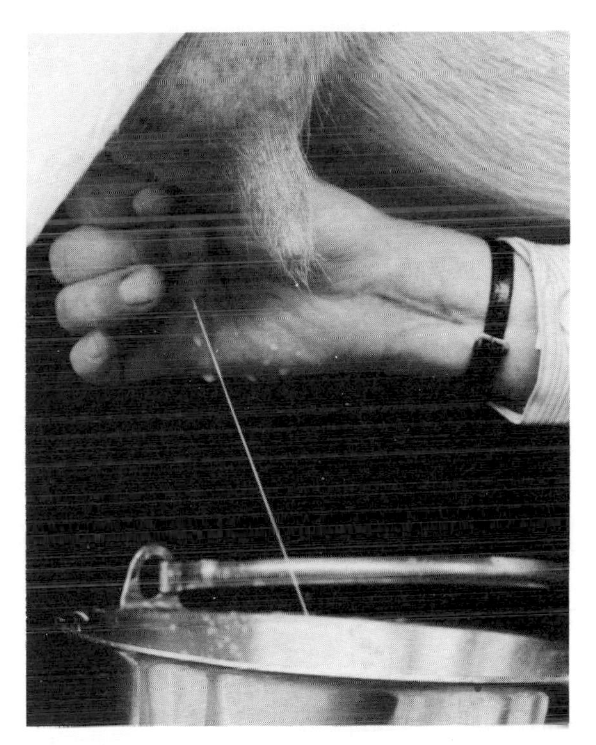

Hand-milking, showing the action of the fingers.

FEEDING

A milking goat giving 4.5 litres (1 gal) of milk a day will require about 2.2 kg (5 lb) of hay per day, as well as green food from browsing outside. The goat is a browser rather than a grazer, and will take hedgerow plants and broad-leaved weeds in preference to grass, although the latter is eaten. She will also need a concentrate ration in the proportion of 15 per cent protein to 65 per cent carbohydrates. A suitable concentrate ration would be made up as follows:

1 part sugar beet pulp
1 part bran
1 part rolled oats
1 part extracted soya bean meal

In spring and summer, the concentrate ration can be reduced because at that time protein levels are higher in green foods than they are in autumn and winter. Ready-mixed goat rations are available from feed suppliers, and many small goatkeepers find it more convenient to buy these, rather than mix their own. Fodder crops, such as those listed on page 46–7 are suitable for goats, but, as their mouths are not adapted to coping with bulky objects, crops such as swedes will need to be chopped for them. Wild plants such as hedge prunings are eagerly taken, as long as they are not poisonous, such as laburnum, yew or privet. Most decorative shrubs are unsuitable, but rose prunings are well received, as long as they have not been sprayed. On the question of poisonous plants, the MAFF publication, 'British Poisonous Plants' (Bulletin 161), is worth reading, while the British Goat Society leaflet, 'Wild Food for Goats', is a useful source of information.

Concentrates are normally fed at milking time, so a daily ration can be split between morning and evening milking. Hay is given in the hay rack in the goats' individual pens, and a convenient time to feed this is immediately after the second milking, when the goats are put to bed. The ration of green food will come from their outdoor browsing and grazing in spring and summer, but in winter this may also need to be put in the pens, in the form of hedge prunings, and winter forage crops. In winter, if weather conditions are such that the goats must remain inside, the hay ration will need to be given in the pen after the first milking in order to keep the goats contented.

CONTROL

Goats destroy hedges and can be difficult to confine. Electric fencing is the most effective and this can be used to control grazing or silage feeding, as in the case of cows. Where electric fencing is not available, and the fences are not secure, it may be necessary to use running tethers, but this does have problems. It takes time to tether the goats, they have to be checked frequently to make sure they have not become entangled, and they need to be moved on to fresh ground every so often. If it starts to rain they cannot get shelter if they are tethered, and goats are not geared to withstand wet conditions and dislike them intensely. Where goats are outside, browsing and grazing, they do need a temporary shelter into which they can run if it starts to rain. This need only be a simple, movable shelter, but is effective if used in conjunction with efficient fencing, such as electric fencing. Clean water should be available at all times, outside as well as inside. If a permanent tank is not available, buckets can be used as long as they are either supported by purpose-made metal supports or placed in old rubber tyres, to stop the goats knocking them over.

BREEDING

Unlike cows, which have a lactation period of nine months, goats can produce milk for up to two years and, therefore, do not need to kid every year. The normal practice amongst goatkeepers, is to have their goats served once every two years, and if this is arranged so that the goats kid one year and the rest the next year, there is an uninterrupted milk yield.

There is, as yet, no artificial insemination service available for goats in Britain, so there is a choice of two courses of action. Either the female must be transported to a nearby male for servicing or a stud male goat will need to be kept on the premises. For this, special housing and an exercise yard will be needed, and he must be kept separate from the other stock. A male does smell, and for those who have never encountered it before, the smell has to be experienced to be believed. Female goats do not smell, but if they are in contact with a male, or his housing they will certainly pick it up. Goatkeepers who keep stud males usually have a special overall which is kept just for visiting him, so that the smell is not imparted elsewhere. Many goatkeepers prefer not to keep a male and either transport their females to a local one, or if there are a number of females involved, they may arrange to have the male brought to them for a few days. Official permission to move goats is not required, but it is necessary to have a 'Movement of Livestock' book and to record the date and details of the movement. If at all possible, it is better to avoid keeping a male. However, if the number of goats kept makes transport difficult, there may be no alternative.

Female goats come into season between September and spring. Once every three weeks or so, they will be on heat for one to three days. The indications of this are a slightly reddish and swollen vulva with possibly a slight discharge. The clearest indication of all is a persistent bleating and vigorous wagging of the tail from side to side. If the female in question is to be transported to a male, there is no time to be lost. A quick telephone call to the breeder is usually all that is necessary to make the appointment. Goats are generally easy to transport, and with few exceptions will stand for the male, so that mating is normally very quick. The breeder will charge for the service and write out a stud certificate. If the service has not taken (indicated by signs of being on heat three weeks later), it is normal to give free servicing to make up for it. Local goatkeeping societies produce a stud list for all their members, so that it is possible to select a suitable male in advance. It is good practice to dust the female with lice and mite powder before she goes to the male, and after she comes back.

Pregnancy lasts for five months, and during this time, the goat is fed a normal maintenance ration with extra concentrates depending upon her level of milk production. Two months before kidding, she should be dried off, so that all her resources can go into keeping herself fit and her growing kids in a healthy state. Drying off is normally straightforward with goats. The technique is gradually to milk less, leaving behind a little milk after each milking so that less milk is produced next time. As the milk yield drops the goat can be milked once a day instead of twice, and then once every two days, until she is finally dry.

For the last two months before kidding, she will need a concentrate ration similar to that of a milker, to allow her body reserves to be built up for her next lactation. This will be about 1.3–1.8 kg (3–4 lb) a day, as well as hay and available greenstuff. This pattern of feeding prior to kidding is generally referred to as 'steaming up'.

Kidding is normally straightforward, but it is wise to have the vet's telephone number to hand, in case of any problems. The udder begins to fill up during the last week or two, and may become so distended that a little milk will have to be drawn off to ease it. On no account, however, should it all be stripped out, for this is the highly nutritious colostrum that is so vital to new kids.

It is usual for goats to have twins, although one or three are not unusual. The normal presentation of a kid during birth is head first, with the nose resting on the front legs. If there are indications of a difficult breech birth, where the kid is the wrong way round, or there is straining for a long time with no result accompanied by obvious distress, then the vet should be called.

The goat will lick and clean the kids herself, and usually they are staggering to their feet quite soon. The sooner they start suckling the better, for the colostrum taken in the first twenty-four hours provides them with antibodies to resist infection until they can produce their own antibodies. As soon as possible, the kids should be checked to establish their sex and whether there are any deformities. Those with deformities should be killed, and if you cannot do it yourself, the vet will do it for you. Males that are being kept for meat should be castrated. Again, the vet will do this, although it is not difficult to do it yourself. A rubber ring is placed around the scrotum, using an elastrator. This piece of equipment holds the ring open to allow it to be passed over the scrotum, then it releases it so that it tightens over the base of the scrotum. It is wise to have the kids disbudded if they have horn buds on the head. There is no reason for a goat to have horns and they are undoubtedly a cause of accidents, even with the most docile of animals. The vet will normally carry this out when the kids are a few days old.

It is wise to register the kids with the British Goat Society if they come from registered parents. Local goat societies will advise on this, as well as arranging an ear-marking session for the season's kids. This is by far the best form of identification, and is both permanent and effective.

After kidding, the goat should be given fresh, warm water with a little salt in it, and plenty of good hay. It is best to withhold concentrates for about two days after kidding, so that milk production is not stimulated to an extent that causes calcium deficiency. This can lead to milk fever, which can be fatal if not diagnosed and treated immediately. A bran mash can be given instead, with concentrates gradually being introduced sparingly from the first day onwards. A mineral lick should be available in the pen. Green food can also be gradually introduced, and after a few days, if the weather is suitable, she can be allowed out to graze. It is a good idea to treat her with a vermifuge at this time, to ensure that she is free of worms. It is normal to allow the kids to feed from the mother for four days, then to bottle feed them from then onwards, so that the milk can be taken for other purposes. The kids will need to be separated from the mother and housed in a communal kid pen, and this seldom creates a problem. Lemonade bottles are ideal for bottle feeding, and both 'plug-in' and 'pull-on' teats are available. Powdered milk is available as a commercial lamb feed, and this is a suitable alternative for kids. Follow the manufacturer's instructions, and ensure that the bottles and teats are sterilized in hot water before use. At first the kids will take about a quarter of a bottle at each feed and will need to be fed four times a day. Gradually increase the amount to half a bottle until by the time they are two or three weeks old, they will be drinking about 2.3 litres

(4 pints) a day. At this time, the number of feeds can be reduced to three, and a little good quality hay introduced. As more bulky foods are consumed, the amount of milk can be reduced, until weaning is complete. When this takes place is a matter of individual choice and can be anywhere between three and six months.

HEALTH
All goats will need worming twice a year, and a choice of vermifuges is available from the vet or from the livestock suppliers. The manufacturer's instructions should be followed, and the milk discarded for three days after dosing.

Enterotoxaemia and tetanus are both killers and it is worth protecting goats against them. This is a simple procedure consisting of injections, which the vet will carry out, and he will also advise on the frequency with which boosters should be given.

Goats, like many animals, are liable to pick up lice and mites, and these can be readily disposed of by dusting with a proprietary insecticide available from the vet. Cuts and wounds need to be cleaned and treated with an antiseptic preparation.

Trimming a goat's foot

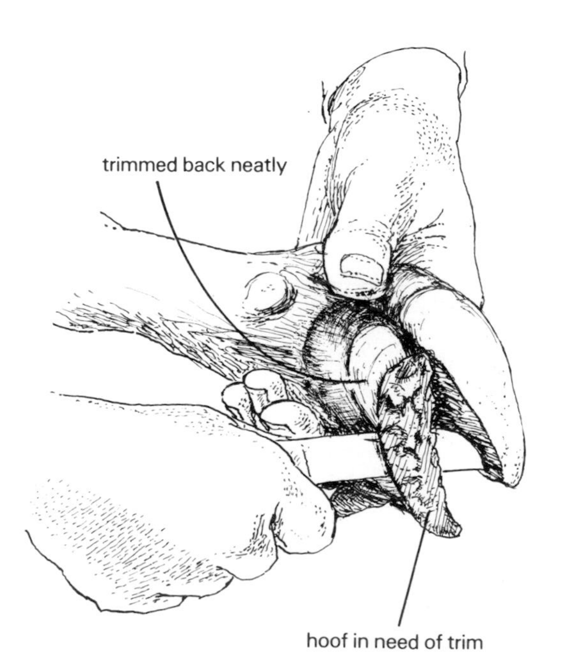

Mastitis can affect goats, and treatment is the same as that detailed in the section on the dairy cow.

Hoof trimming will be necessary, about once every six weeks, as the claws of a cloven hoof grow continuously. Trimming involves clipping back the claws which otherwise grow and fold inwards over the hoof, and is not difficult to carry out. Once trimmed level, the hoof is smoothed with a rasp.

Milk sheep

Sheep which are kept for their milk-producing ability are mainly to be found in the Middle East and Mediterranean areas of the world, and are characterized by having longer legs for easier access to the udder. Some breeds in fact look more like goats than sheep. The most important of the breeds is the Fries Melkschaap, which is found in Holland and Northern Germany, and which has also been imported into Britain. It has the highest milk yield, averaging 636 litres (140 gal) in a 250-day lactation. The milk has a butterfat content of 6–7 per cent which tends to stay in suspension like that of goat's milk, rather than immediately separating and floating to the surface. It is a rather delicate breed, and does better in small flocks where a greater degree of individual attention is possible. The feet are tender and liable to foot rot, so well-drained pasture is necessary, as well as adequate shelter throughout the year. They relish fairly coarse grasses, as well as hedgerow pickings, and do well on a concentrate ration as long as there is sufficient roughage to balance it in the diet. Hay and barley straw are ideal.

The Fries Melkschaap is not an ideal breed for meat production and is better crossed with another breed for this purpose. A Southdown or Dorset Down ram is an excellent choice, producing good meat lambs. The Lacuna or Lacaune is the premier milk breed in France, whose milk is used to produce the famous Rocquefort cheese. It is more numerous than the Fries Melkschaap but has a lower yield. The Larzac is also found in France. Other milk

Fries Melkschaap milk sheep.

breeds found in different parts of the world are the Pramenka in Yugoslavia, the Daglic and Sakiz in Turkey, the Awassi in Israel and the Middle East, the Chios in Greece, the Sopravissina in Italy and the Barbary in the general Mediterranean area. For someone who wishes to establish a small dairy flock and to obtain sufficient yield to make yoghurt or soft cheese production viable, the breed to choose is the Fries Melkschaap, or hardier hybrids developed from it.

HOUSING

Housing can be simple, the important factors being shelter from wind and rain, dry flooring and good ventilation. Poor ventilation can lead to problems such as pneumonia and other respiratory complaints. A structure which is boarded to a height of 1.5 m (5 ft) will provide shelter while a space between that and the roof will ensure adequate ventilation. On the north- and east-facing sides, it is as well to board the sides entirely to cut out winds, rain and snow. Straw provides bedding within the house and this is added to in order to maintain

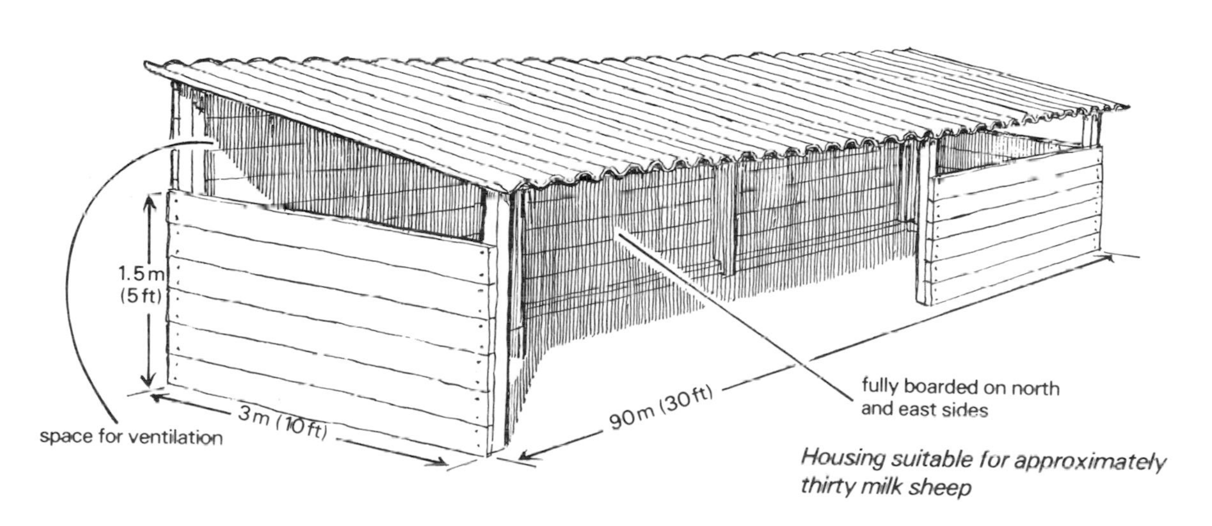

Housing suitable for approximately thirty milk sheep

clean conditions. It should be cleaned out regularly so that a damp environment, which can lead to foot rot, is avoided.

When sheep have just lambed, it is best to provide individual pens for each with her lambs. This is easily arranged by utilizing movable sheep pens and placing them either on one side of the existing house or, ideally, in a second similar house.

Drinking buckets in metal stands, such as those provided for goats, are adequate; the important points to ensure are that they cannot be knocked over, and are high enough to avoid contamination by dung, yet low enough for access. Approximately 60 cm (2 ft) above the ground is a suitable height. Hay racks should be available in the house, to provide access to hay, feeding straw or winter feed prunings such as willow branches. Troughs for the feeding of concentrates need to be no higher than 60 cm (2 ft), and each sheep will need approximately 45 cm (1 ft 6 in) of trough space. Ideally they should be fixed to the wall so that they are not knocked over or contaminated by droppings.

MILKING

The milking process is the same as that detailed for cows and goats, with similar emphasis on hygiene and clean milk production. Machine milking units are available for sheep, and this is obviously ideal, as long as sufficient capital is available for the purchase of equipment. Hand-milking is more difficult with sheep than with goats because the teats are shorter, and it is a more laborious process. A milking platform is essential otherwise the strain on the back of the milker is considerable.

FEEDING

In spring and summer the sheep will get most of their nutrients from grass, but in spring when the new grass is lush, it is a good idea to give them a hay or barley straw ration first to avoid scouring. They will need a concentrate ration such as that given to goats if they are also in milk production, and this will vary depending upon the yield. Somewhere between 450 g and 1.3 kg (1–3 lb) a day is adequate, as long as grass and hay are also being given. A mineral lick in the house will ensure against mineral deficiency diseases, such as swayback resulting from a lack of copper. In autumn and winter more hay and barley straw will be given to make up for the lack of grass. Silage is also readily accepted.

BREEDING AND HEALTH CARE

Details of breeding and health care are given in the general chapter on sheep.

7 DAIRY PRODUCTS

There are six marketable dairy products – milk, cream, butter, yoghurt, soft cheese and hard cheese. The choice of milk lies between cow, goat and milk sheep, but as some milk is better for certain products than others, it is worth examining the options in more detail. If fresh milk is to be sold, then the choice would be either cow or goat, depending upon the situation and interest of the individual. The level of production of cow's milk would need to involve at least twenty cows, as any enterprise smaller than this would make it uneconomic in terms of capital expenditure on equipment to meet dairying regulations. A goat enterprise could be much smaller, for the marketing is carried out by the producer, and is not subject to the same regulatory structure as that involving cows. There are many goat keepers who have fewer than six goats, and who sell surplus milk to local consumers and retailers.

It is not economic to sell sheep's milk, because not only is there virtually no demand for it, but also the labour costs per pint are higher than those for a cow or goat. Because of this, it is much better to concentrate on using it to produce a marketable commodity that will command a higher price, such as yoghurt or soft cheese. If the emphasis is to be on cream sales, the only animals worth considering are cows, with the Channel Island breeds, Jersey and Guernsey or the South Devon, taking priority. Goat's milk is naturally homogenized, which makes cream separation more difficult, and the butterfat content is lower than in cow's milk. Although sheep's milk has a relatively high butterfat level, the lack of demand would not warrant trying to make cream from it.

Although butter can be made from the cream of any milk, it is only worth making it from cow's milk. The cream is easy to separate and buttermaking is simple on a small scale. However, it is not economic in terms of time, or return, to try and sell butter on a part-time basis. The only viable method of production is the large-scale, fully commercial and automated enterprise. Anyone who has ever made butter will know that the time goes not on making the butter itself, but on washing and working it to remove surplus water. Where small quantities are being produced for family use, this poses no problem, but with a large amount it is difficult to wash the butter and dispose of the surplus moisture effectively. Butter intended for sale should not contain more than 16 per cent of water in order to comply with regulations.

Yoghurt is a good source of revenue for those who are geared to producing it. Most of the yoghurt sold in supermarkets contains artificial colourings, flavourings and preservatives, but there is a growing market for unadulterated products. Small farmers already equipped with buildings and equipment for a small commercial herd of cows, may find that the return on a product like this is better than that from liquid milk sales. It was the late Dr Schumacher (author of *Small is Beautiful*) who said that any enterprise that is a prime producer should aim to get the 'value added' by turning the product into something with a higher value. Cow's milk turned into yoghurt may lose the security of the MMB monthly cheque, but, as long as the distribution is organized, it will provide a higher return. If cream is already being sold, the skimmed milk can be used to make yoghurt, so the two activities dovetail together well. Goat's milk makes excellent yoghurt, and has a particular appeal for those who suffer from allergies and who normally cannot take cow's milk products. Sales outlets are generally through wholefood and delicatessen shops.

Sheep's milk, with its high butterfat content, makes good yoghurt, and should appeal to the luxury food market, but the main obstacle to be overcome is lack of knowledge on the part of the public. Soft cheese made from cow's milk is widely available, but that made from goat's and sheep's milk is a rarity. This is a pity for there is a good potential for both. We have only to look at the range of cheeses available in France, Italy and other European countries, to see how both goat's and sheep's milk can be utilized.

Making a hard cheese is much more time-consuming than making a soft one. Apart from the initial production and pressing time, there is the ripening and storage period, during which space and capital are tied up. It is, therefore, only appropriate to a fairly large-scale operation. Cows, goats and sheep's milk can all be used to make a range of hard cheeses, but they are best kept for family use, unless a full-scale and large enterprise is envisaged.

Regulations naturally apply to the production of dairy products, with the most stringent applying to those made from cow's milk. Anyone wishing to sell cow's milk products must apply for registration to the local MAFF.

Registration is not necessary for goat's milk products, but normal health regulations in relation to any food production will apply. The production of dairy products from sheep's milk is still new in Britain with few commercial flocks in existence. It is best to apply direct to the local MAFF for their ruling, because, at the time of writing, there is a certain amount of official confusion as to what regulations apply to a sheep that is producing milk.

The table below gives the salient points about each dairy product in relation to the milk source. It is, at best, a basic guide, for local conditions will prove to be major factors for anyone trying to make a decision about which dairy products to specialize in.

The dairy

A commercial enterprise which involves the production of dairy goods will necessitate a properly equipped dairy to meet health regulations. If small amounts of goat's milk products are being produced for sale, a normal kitchen may suffice, but production will be curtailed when other activities have to be conducted in the same area. Yeasts associated with bread making and fruit preserving can have an adverse effect on cheese and yoghurt.

One of the most important aspects in a dairy is cleanliness, for there are many bacteria and moulds which find milk an ideal medium in which to grow. The ceiling and walls should be dust free and non-flaking, while the floor should be stone flagged or rendered concrete so that regular hosing down with water is possible. A drain will be needed to take away surplus water.

Windows let in light, but, unless they are kept closed, they also let in unwelcome pests such as flies. The ideal is to have an air extractor fan for ventilation.

Both hot and cold water will be needed, the former for cleaning surfaces and equipment and

Which dairy products?

Dairy animal	Milk	Cream	Butter
Cow	Regular milk cheque from MMB for sales of liquid milk, but must be registered, and buildings, equipment and practice must comply with regulations. Small enterprise not viable unless existing buildings can be used.	Good sales of cream possible, but must be registered, and comply with regulations. Best breeds are Channel Islands, Jersey or Guernsey. Skimmed milk can be utilized with powdered milk to produce yoghurt.	Not commercially viable on a small scale, but worth doing it for family use.
Goat	Sales to local consumers and retailers possible. No registration required.	Not commercially viable, but worth doing for family use.	Not commercially viable, and generally not worth doing for family use unless goats have particularly high butterfat levels.
Sheep	Not commercially viable	Not commercially viable	Not commercially viable

Which dairy products? cont.

Dairy animal	Yoghurt	Soft cheese	Hard cheese
Cow	Good sales possible, and dovetails well with cream production enterprise. Must be registered and comply with regulations.	Good sales possible of cream and lactic cheeses. Must be registered and comply with regulations.	Not commercially viable on a small scale, but worth doing for family use.
Goat	Good sales to local consumers and retailers. No registration required. A better return than on liquid milk.	Good potential in a market that is almost entirely unexploited. No registration required.	Not commercially viable, but worth doing for family use.
Sheep	Good potential for delicatessen and luxury food sales.	Good potential for delicatessen and luxury food sales.	Not commercially viable, but worth doing for family use.

the latter for hosing down the floor. A small Burco boiler is ideal as a source of hot water, and also allows equipment to be sterilized in it. Working surfaces should be easily cleaned. Traditionally, slate or marble were used, but stainless steel or formica is adequate, if less cool.

Electric points, preferably waterproof, for a boiler, cooker or any other equipment should be placed high up, well away from any water which may splash during hosing down the floor. Appliances such as a cooker should be on supports, to lift them clear of the floor. The other safety factors to bear in mind, are that heavy equipment such as centrifugal cream separators should be bolted down to a working surface and any knives or dairy chemicals stored well out of reach of children. No animals should ever be allowed into a dairy.

Milk

As already indicated, cow's milk can only be sold through the Milk Marketing Board, and the farmer must be registered, following an inspection of premises, equipment and techniques. Collection of milk is by bulk milk tanker once a day, and the farmer receives a monthly cheque for his milk. This cheque does provide a certain security although the smaller the enterprise, the less cost effective it is. The farmer with fifty cows will obviously have a higher rate of profitability than the one with twenty cows. With a pipe-line milking unit, it does not take much longer to milk fifty cows than it does twenty. The small farmer may well find that it is better to use his milk for a product with a higher premium such as yoghurt.

Goat's milk has a ready market in wholefood

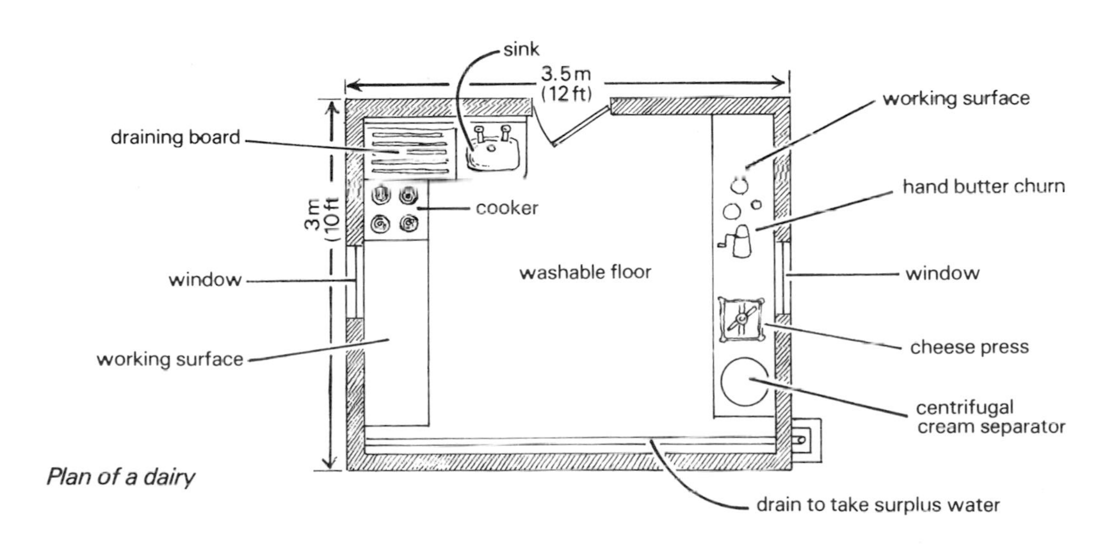

Plan of a dairy

shops, and most of the milk sold in this way is produced by small herds, often of six or less goats. There is no state help or advice in relation to the distribution of goat's milk, and it is something which is left entirely to the individual goatkeeper. There are moves afoot to start a marketing co-operative in Wales, at the time of writing. In parts of the USA, there are large co-operatives distributing and selling goat's milk on behalf of their members.

In Britain there are two ways in which goat's milk is packaged for sale, either by using waxed cartons or purpose-made plastic bags. Both are satisfactory, and can be stored in the deep freeze, but the packaging does represent a considerable outlay for the small goatkeeper. The best way of buying cartons or plastic bags is in bulk, for there is normally a discount for bulk buying. Goatkeeping societies frequently buy in bulk and it is worth trying to get as many individuals to place a combined order through such a society in order to have the benefit of a bulk discount.

Waxed cartons are usually pre-printed with a picture of a goat, a description of the milk and the volume contained. They are closed by means of metal sealing clips, for which a bench-top clip-sealer is available.

The plastic bags are about half the price of cartons. They are also pre-printed with an attractive logo and a panel for the individual goatkeeper to insert a label showing his own name and address. The method of sealing the bags is to use a heat-sealer such as is commonly used for freezer bags. An electric sealer is available quite cheaply for this purpose. Specially shaped rigid containers are also available for the handling of the bags, which would otherwise be difficult to pour without spilling.

Sheep's milk is still such a rarity, that it is unlikely there would be a demand for it. As it has a high butterfat content, it is not as palatable in texture as cow's or goat's milk, and is more suited for yoghurt or soft cheese production.

It is worth emphasizing again, that the best return on any milk is when you can turn it into a higher premium product yourself, and if you are already in a milk production enterprise, it is worth giving this some consideration.

Cream

When milk is left to stand, the cream rises to the surface and forms a layer on the top. Even goat's milk, which is naturally homogenized, and where the smaller fat particles are evenly distributed throughout, has a tendency to do this. Traditionally, the cream was skimmed off with a purpose-made skimmer or perforated ladle, and for small-scale cream production this is suitable. If commercial production is envisaged, however, a centrifugal cream separator is essential. This separates the cream in the most efficient way, ensuring that no cream is left behind, as it is with even the most careful hand skimming.

For cream production, only cow's milk is really practicable, with that of the Jersey, Guernsey, North Devon and South Devon breeds taking precedence. These breeds have a higher butterfat content to their milk than other breeds, but this is not to say that the milk from breeds such as Friesians cannot be used – there is just less cream.

No cream from cow's milk may be sold without a licence, and again, contacting the local MAFF office is the first step to achieving this. Premises, equipment, water supply and dairy practice will all be subject to inspection and samples of milk and cream will be taken regularly. It is obvious, therefore, that selling their own cream is practicable only for those who are already geared to keeping a dairy herd and who have suitable conditions.

The type of cream sold is subject to regulations, for there is a considerable variation in the amount of butterfat, depending on the speed at which the cream separator was going. The amount of butterfat can be regulated by adjusting the speed of the separator. The types of cream are as follows:

Clotted cream	55 per cent butterfat
Double cream	48 per cent butterfat
Whipping cream	35 per cent butterfat
Sterilized cream	23 per cent butterfat
Single cream	18 per cent butterfat
Half cream	12 per cent butterfat

The higher the percentage of butterfat, the

greater the price of the cream. Plastic pots with snap-on lids are available, as well as equipment for pressing on embossed foil tops. The latter, however, is better suited to large-scale production, for the cost of such equipment is high. For the small producer, pots with snap-on lids are quite suitable. The cream from Jersey cows can be put in special pots which are produced by the Jersey Society. These are pre-printed to indicate that the cream is Jersey and the quantity is shown.

Yoghurt

Yoghurt is essentially milk which has been curdled by the action of acids from *lactobacillus* organisms on milk protein. There are several types of suitable bacilli, but the ones which are used as a 'starter' for yoghurt are *Lactobacillus bulgaricus*, *L. acidophilus* and *Streptococcus thermophilus*. Yoghurt can be made from cow's, goat's or sheep's milk, and the process is the same. It can also be made on any scale, using anything from a Thermos flask, to a commercial vat. For a small, commercial enterprise, production in relatively large quantities is obviously preferable. Several large, family-sized Thermos flasks can be used to make a batch of yoghurt, and this is transferred to individual pots when ready. If the scale of operations warrants it, a small vat is available, and this simplifies the process.

Milk is heated to 82°C (180°F), then cooled to 43°C (110°F). A dairy thermometer is essential, because if the temperature is too high when the 'starter' is added the the bacilli will be killed. Alternatively, if the milk is too cold, the metabolic action of the organisms will be slowed down.

For commercial yoghurt production, a commercial starter is essential. It is possible to use some shop-bought plain yoghurt as a source of *lactobacillus* organisms, but there is no guarantee that they are sufficiently strong to coagulate

Above right: A cream separator in action.

Right: A yoghurt vat makes small-scale commercial production easier.

the milk, and you may end up with a batch of watery, insipid-looking yoghurt. Commercial starters are generally available in freeze-dried sachets, which can be stored in the deep freeze for up to six months before use. To make up the starter, the manufacturer's instructions should be followed closely. This involves adding the culture to 1.5 litres (2¾ pt) of pasteurized and cooled milk that has been brought to 43°C (109°F), and then left to incubate overnight, until needed. During the incubation period, it should be covered to keep out dust. If a large quantity of yoghurt is being made, all the culture can be used, otherwise some can be frozen for future use. The quantity of starter needed is 5 ml (1 teaspoon) to every 0.5 litres (1 pt) of milk. This is stirred in, and the whole lot is then left in a temperature of 43°C (109°F) until coagulation is achieved. The yoghurt is then transferred to individual plastic pots which have been sterilized in hot water, and the snap-on lids then provide an effective seal. The pots are refrigerated until needed, and the quicker it is sold the better, for the longer it is kept, the more acid it becomes.

If fruit yoghurt is to be produced, the solid fruit must be added after incubation, otherwise it sinks to the bottom. Blanched and cooled raspberries or strawberries are excellent for adding to yoghurt, and sugar or honey may be added to taste. The pots must be labelled with the contents, in order to meet the trade regulations; for example, *Contents: Goat's milk, Lactobacillus bulgaricus culture, raspberries, sugar*. The quantity of yoghurt should also be clearly marked.

If cow's milk is being used, the skimmed milk left behind after cream separation can be used, but dried milk will be needed to provide bulk for the yoghurt, and this represents an added cost. However, there is a higher profit to be made from removing the cream and selling that separately, and then adding milk powder to the yoghurt, than there is in making yoghurt from whole milk. It is, of course, necessary to include the dried milk in the list of contents on the pot. It is not a good idea to use dried milk in goat's milk yoghurt, for many people buy goat dairy products for health or allergy reasons, and the inclusion of dried cow's milk would be unacceptable. As yet, there is no dried goat's milk available commercially in Britain.

Soft cheese

The manufacture of soft cheese can produce a good return for those with an existing milk source. Most cheese available is made from cow's milk, but there is undoubtedly an unexploited market for goat's and sheep's milk cheese. This market embraces the health and wholefood as well as the delicatessen and luxury food areas.

Cheesemaking is governed in Britain by the Milk and Dairies (General) Regulations 1959, The Cheese Regulations 1970, The Food and Drugs Act 1955, and by a variety of other legislation covering labelling and the use of additives. Copies of these are available through the HMSO bookshop. It is also worth contacting the local MAFF office, to ask the Dairy Officer for his advice, for he is in a better position than anyone to give information and help in relation to local conditions.

Soft cheeses must contain a certain proportion of milk fat before they can be given specific descriptions. These are as follows:

Full fat soft cheese	not less than 20 per cent milk fat and not more than 60 per cent water.
Medium fat soft cheese	between 10–20 per cent milk fat and not more than 70 per cent water.
Low fat soft cheese	2–10 per cent milk fat and not more than 80 per cent water.
Skimmed milk soft cheese	less than 2 per cent milk fat and not more than 80 per cent water.

If all this seems rather confusing to the small cheesemaker who may not be in a position to take regular milk fat samples, and who just wants to sell some goats' cheeses, all is not lost. Talk to the MAFF Dairy Office, or contact the local authority and make it clear what it is you want to do. There are specified circumstances where the specific labelling requirements are not necessary, and this is largely determined on the basis of individual and local circumstances.

There are many different recipes for soft cheeses, and it is a matter of individual preference which ones are selected. As a general rule, goat's milk produces a slightly softer curd than cow's milk; apart from this, the recipes can be used equally well for all kinds of milk. A specialist book on home dairying will give comprehensive details of individual recipes, but a good general purpose recipe is as follows:

Heat the milk to 68°C (155°F) and cool immediately to 32°C (90°F). How this is done will depend on the volume of milk used. For large quantities, a milk churn placed in a boiler will suffice, while smaller quantities can be dealt with in a double boiler or large pan. A Burco boiler is particularly useful as it can be used for the hot water sterilization of equipment as well as for heating milk.

Cooling is effected by placing the milk container in cold water. A deep sink or tank on the floor is suitable. When the temperature of the milk is at 32°C (90°F), a commercial cheese starter culture is added, at the rate of 5 ml (1 teaspoon) to every 3.5 litres (6 pt) of milk. The addition of this provides the appropriate bacilli, giving the cheese its appropriate 'aroma'. Two hours after adding the starter, add cheese rennet in the proportion of 5 ml (1 teaspoonful) to every 3.5 litres (6 pt) of milk, and leave the milk to coagulate for one hour. It will be ready when the back of the finger is pressed on the surface, and no milk stain is apparent on the skin as it leaves the curd. Using a long-handled blade, separate the curd from the sides of the container, and cut first one way, then the other, to make it easier to remove. Prepare stainless steel buckets lined with butter muslin draining cloths and pour the curds and whey into them. The curds will be retained by the cloths, while the whey runs into the buckets. Meanwhile prepare individual moulds and cheese mats by sterilizing them in hot water. The small plastic moulds are open at both ends and are available from suppliers. They need to be used in conjunction with mats.

The mats are laid on plastic trays, then the moulds placed on the mats. The cheese curd is then ladled from the cloths into the moulds pressing down slightly to make sure that no large air spaces are left. The moulds are filled, then covered with mats and left to drain overnight. The following day, scald the top mats to sterilize them, and turn the cheeses upside down onto them. By this time, they will have shrunk to half their original height and will be firm enough to handle, and to take out of the mould. Sprinkle salt onto the surfaces and leave for a further twenty-four hours. They are now ready for sale, and can be packaged in foil.

8 BEEF CATTLE

A large proportion of beef comes as a spin-off from the dairy industry. Friesian cows are crossed with a beef breed such as Aberdeen Angus or Hereford so that the calves are suitable for rearing as beef cattle. Dairy cattle, however, need good pasture, and not all areas have suitable grass. In regions where pasture is poorer, there may be more emphasis on purely beef enterprises. Beef cattle will do reasonably well on poorer pasture and are often fed some straw as a cheap form of roughage. They are frequently found in arable areas where straw is cheap and plentiful, and where the pasture is not of a high quality.

In Britain there is a system of compensating payments for those breeding and rearing cattle in less favoured farming areas such as hill farming regions. Further details can be obtained from the divisional office of the MAFF.

Depending on the situation, a small enterprise may therefore be part of an existing dairying activity. A part-time farmer who has no dairy animals but a few spare fields could consider buying in some beef calves and raising them on his grass. Turning grass into beef is one of the best ways of utilizing the fields. The fences need to be strong, as bullocks have a tendency to lean against them. Further details will be found in the section on fencing. The easiest way of confining frisky beef cattle when they need to be handled, such as for administering some form of medication, is to hold them between two gates. This is arranged by having one gatepost with two gates attached to it.

The dairy farmer will be producing his own calves on a regular basis to ensure that lactation continues in the herd. The small farmer without dairy animals will have to buy calves in from elsewhere. It is highly unlikely that he will breed his own beef cattle as this is a specialized activity, needing considerable capital. It is far better to buy calves and raise them to killing age. When this will be depends on the particular market being catered for, because there are different types of beef, depending on the kind of animal.

Veal is produced from calves fed on milk replacement substitutes and killed any time between four and sixteen weeks of age. The intensive method of confining calves in dark, hot boxes on slatted floors has produced a public reaction to the extent that many people now refuse on principle to eat veal. Baby beef comes from calves that have been weaned, then fattened on barley. Prime beef comes from beef and dairy crosses, when the heifers are killed at eighteen months old, and the bullocks at about twenty to twenty-four months old. Mature beef comes from three-year-old beef cattle, but this is now comparatively rare. Fat cows, or dairy cows which are no longer productive also produce beef, and the most important source is the Friesian cow.

For the small, part-time farmer, the choice will be between baby beef where he is rearing calves up to the age of twelve to fourteen months, and prime cattle, where they are raised to eighteen or twenty to twenty-four months depending on whether they are heifers or bullocks.

The dairy farmer can either sell his calves straight after weaning or raise them to baby beef or prime beef stages. He may also use some of his cows as suckler cows, so that artificial feeding of the calves is unnecessary. These cows are not part of the dairy herd but are kept specifically for breeding and rearing calves. A particularly good cow can be used for multi-suckling, when she rears batches of calves in succession. Calves reared in this way do extremely well, but a careful eye needs to be kept on the cow. If a calf is tending to suck only on one side, there may be an accumulation of milk in the other quarters, and this will need to be milked out.

Raising beef calves

The young calves will need warm, dry quarters, well supplied with bedding straw. The other necessities are a hayrack, a drinker and a trough or feed container for concentrates. If not already weaned, the procedure outlined in the dairy cow chapter for bucket feeding should be followed (see page 79). They should also be castrated, otherwise they will be aggressive and difficult to manage.

Once weaning has taken place, and the calves are feeding on a weaner's ration and hay, they can go out to graze if the weather is mild. This is the point at which problems are likely to occur, for young, fresh grass can quickly lead to scouring. Before they go out, they should be given a hay or concentrate ration first, and their grazing should be strictly controlled. An hour or two is quite enough for the first few days, and then they are brought in again. Once they are used to the change of diet, their grazing can be controlled by electric fencing. Many people find that it is better to pen them in an exercise yard with an attached shelter at first, so that they become used to being out gradually. They may also be given a little green food, such as kale, to accustom the digestive system to greens before they go out to grass. Whether the calf needs concentrates at grass depends on the weight of the calf, the amount of grass available and time of year.

As the calves grow, and the weather becomes warm, they will stay outside, although some form of shelter is a good idea until they are really hardy.

In winter, there is no point in the cattle staying out, for the grass is no longer growing

A consignment of beef calves at the market.

During the summer months, check stock regularly for signs of lice, mite and tick attack.

and is likely to be damaged by the treading action of feet. One way of keeping them during this period, is to yard them. This is an enclosed yard which acts as an exercise area, with a roof shelter, either covering the whole yard, or part of it. It is similar to the system suggested for dairy cows (page 73).

The yard is kept covered with wheat straw, which is added to as necessary, and as this becomes trodden in, it provides an excellent source of manure for subsequent use with crops. Troughs for feeding, as well as drinkers, should be available, and as the cattle are more confined in these conditions than they are on grass, the value of debudding the young calves becomes apparent. Quite serious accidents can occur with horned animals. Silage, kale and root crops, and concentrates such as barley and oats can all be fed in the troughs, while hay is made available in racks. When barley straw is given, it can be put on the ground, as a bale (but remembering to cut and remove the twine). In this way, the cattle are less likely to get the needle-like awns of barley seeds in their eyes, which they might if they were reaching up to take the straw out of racks. Oat straw is the best form of feed straw, but this is not available in every area, and barley straw is easier to buy.

Intensive fattening, which is commonly practised by large beef farmers, utilizes the so-called barley-beef system. This is a method of restricting the hay and roughage intake of the animal and making concentrate feed available on an *ad lib.* basis, so that the barley concentrate is taken in at the expense of roughage. In this situation, the animals, in their search for roughage, will try to eat their bedding straw. Obviously, this must be prevented in some way, such as driving the cattle around their pen while new straw is being added, so that it is trodden down and soiled immediately. This makes it unappetizing to them. Another method of fattening is by hormone implantation. A tablet is implanted in the loose skin behind the head. This system is regarded with dismay by many who are not only concerned about the welfare of livestock, but also by the long-term effect on people, particularly children, who are eating beef with hormone residues.

Selling beef cattle

The main sales of beef cattle will be either through local farmers' markets, or through the auspices of the Fatstock Marketing Corporation. Further details of these are given in the section on selling pigs (page 109).

Local butchers may be interested in locally produced beef and will buy 'on-the-hoof' for slaughtering in a registered slaughterhouse.

9 PIGS

In recent years, the keeping of pigs, like most other aspects of farming, has become big business, with increasing emphasis being placed on intensive factory production. Housing has become so sophisticated that lighting, ventilation, slurry disposal and stocking ratios are all rigidly controlled. Sows are frequently kept in individual stalls which do not allow them to turn around, so that the dung falls through slats in the floor behind them and thence into a slurry channel. When they are about to farrow, or give birth, they are placed in farrowing crates that allow virtually no movement. The piglets, after three weeks, are put in a weaner house with slatted floors for the dung to fall through, and are later transferred to a fattening house. In these environmentally controlled houses, the light is extremely dim in order to keep the overcrowded pigs docile, and to prevent them fighting and damaging each other. Their tails are docked so that they do not chew them and high levels of antibiotics are administered to counteract infection. It is no wonder that there is increasing public reaction against such practices.

The small, part-time farmer does not need to emulate these activities, nor does he necessarily need to base his practices on outdated methods. The sensible approach is to use the benefits of modern veterinary research, to protect his pigs against disease and parasites by administering vaccinations and vermifuges, and yet respect the natural instincts of the animal by giving housing and conditions that are as natural as possible. Anyone who has ever seen the immense care with which a pregnant sow makes her nest, carrying straw in her mouth and placing it carefully, would find the cold, strawless, metal farrowing crate an abomination.

Pigs are highly intelligent and sensitive creatures. It has been said that they are the most intelligent livestock on the farm. They are also naturally clean and hardy animals, and as long as they are given sound weather-proof and draught-proof housing, clean conditions and adequate food and water, they will do well, without falling victim to the diseases of the intensive house.

There are several stages at which pigs can be sold off. They may be bred and sold as weaners at about five to eight weeks old, or they may be kept on and sold as store pigs at the age of twelve weeks. Porkers are pigs sold for fresh meat, and they may be lightweight, weighing 45–68 kg (100–150 lb), or heavy porkers (cutters), weighing 68–90 kg (150–200 lb). If the pigs are kept on after this they are sold as baconers from about six months of age, when they will have an average liveweight of 72–90 kg (160–200 lb). Baconers provide pig meat for factory curing, which produces either 'green' or 'smoked' bacon.

A good way for the small farmer to start with pigs is to buy in weaners and raise them to pork or bacon weight before selling. This will not involve him in expensive housing, and it also means that it is simple to work out what the feeding costs will be beforehand. A good conversion rate for a pig would be 3.5, which means that it needs 1.5 kg ($3\frac{1}{2}$ lb) of feed to produce 500 g (1 lb) of liveweight. The growth rate to aim for would be a 45 kg (100 lb) porker in four months, a 90 kg (200 lb) porker in six months and a heavy pig in seven months. A bought-in weaner at the age of six to eight weeks would weigh 15–18 kg (35–40 lb). In recent years the practice has been to wean earlier so that weaners are available as early as three weeks old.

If the farmer wants to breed his own pigs, once he has gained experience with weaners, an

Above: The Landrace pig produces more lean meat than some of the older, more traditional breeds.

Below: The Gloucester Old Spot, a traditional and hardy breed that is popular with small-scale farmers.

inexpensive way of doing this on a small scale is to select a few of his purchased weaners that look as though they will make good mothers, and keep them for breeding. The characteristics of a potentially good sow are a good, healthy growth, twelve to fourteen well-formed teats and a docile, gentle nature. The latter is a characteristic that is much more important to the small farmer than it is to the factory pig producer. The small farmer is more likely to know his pigs individually, and generally has a better relationship with them. Further details of breeding are given in the appropriate section.

No pigs can be moved to and from premises without a permit from the Inspector of Diseases of the local authority. This is in addition to the normal recording of movements in a record book kept for the purpose.

Where waste food is given to pigs, the processing of it must conform to the Diseases of Animals (Waste Food) Act 1973. This requires that the waste food is boiled in water for at least an hour, in specially designated premises, where treated and untreated food are kept quite separate. The object of these regulations is to eliminate the possibility of disease being recycled. There are similar regulations in the United States, but in parts of Australia all feeding of swill is banned.

Breeds

The decision as to which pig breed to choose is a personal one. The modern hybrids based on the Large White and Landrace breeds are long and lean, so that the proportion of fat to meat is less than in older types. They are quick-growing and are certainly the best choice if a quick turnover is aimed for. The main criticism of some of the hybrid strains is that, because they have been bred for intensive, indoor conditions, they may not be as hardy as some of the more traditional breeds.

The Welsh and the British Saddleback are both hardy and are good for crossbreeding, while the older breeds Tamworth and Gloucester Old Spot are particularly popular with smallholders who keep hardy outdoor pigs. In the United States, the Hampshire and Duroc are frequently kept.

For those whose main interest is in the breeding and perpetuation of old breeds, the Rare Breeds Survival Trust is an organization set up with these objectives in mind. In addition to pigs, the society covers other livestock, including, cattle, sheep, goats and poultry.

Housing

The small, part-time farmer, whose aim is to produce good-quality pigs in an economic yet humane way, will need a building to house the weaners when he first takes delivery of them. In addition, if they are being outdoor-reared in the summer months, they will need a movable house in the field. If a farmer is breeding his own stock, he will need a farrowing house for the sow to give birth.

The buildings must be soundly constructed and provide warmth and freedom from draughts and damp. Ideally there should be an attached yard for exercise and dunging, and the fence for this will need to be strong. The traditional stone walls are excellent, and these usually had a feeding trough built in with a gap above it so that feed could be introduced from outside. Pig netting is effective as long as it is firmly attached to stout timber or metal posts. The base is particularly vulnerable, for pigs get their noses underneath netting and force it up. They are far stronger and more crafty than many people realize. A stout base plank or metal rail is necessary, although the former may need periodic replacement if it is chewed. Small farmers have also used galvanized, corrugated iron sheets attached to a stout timber frame for yard walls.

The ideal yard slopes slightly away from the house, and has a draining channel on one side. The aim is to remove the dung every day, and usually this is in one place so it is not difficult. Once a week the yard can be washed down with a hosepipe so that it drains into the draining channel. With this practice, there is no slurry problem – a major problem with the large pig producer. The dung which is taken out every day is relatively dry and can be stacked for

Portable shelters are useful for pigs kept outside.

composting. Pig manure is highly concentrated, and once rotted down, is one of the best natural fertilizers available.

The house itself should have a strong floor, and concrete or stone is best, otherwise the pigs root about and dig up rammed earth. The floor must be kept covered with straw for warmth; the pigs will take great trouble to rearrange the straw in order to make a bed.

A pig trough will be needed for food, and this differs from any other trough in being made low for easy access, and heavy based for added stability. Galvanized metal is best, for any other material would either tarnish or would be chewed up. A stout, galvanized metal water trough is the only other necessity. Pigs, like all livestock, need access to fresh, clean water at all times. In larger units this will be laid on automatically, and it does save a lot of time. Feeders and drinkers should be thoroughly cleaned and washed regularly. The principle that we would not drink out of a dirty cup is just as valid when applied to the pig and her drinker.

When pigs are on pasture during the summer months, one acre for every eight to ten pigs will be needed, depending on their size. There should be field shelters and the traditional type was a movable ark. They need only be simple structures as long as they provide warmth and dry conditions. One of the best is a galvanized shelter with a rounded roof and built-in floor. Straw is placed inside the shelter and the whole structure can be moved when the pigs are transferred to new pastures.

Pigs on grass

Pigs on grass are more liable to pick up worms than those that are indoor-reared. It is important to worm them frequently with a vermifuge prescribed by the vet. The pasture should be fresh and unused by pigs the previous season.

One of the most difficult aspects of keeping pigs out of doors is confining them. In areas where adequate strong fencing is not available, and for controlled grazing, electric fencing can be used to good effect.

Traditionally, the pig was used as a natural plough, and there is certainly no livestock as effective at rooting up and generally turning over the soil. For this to happen effectively, the pigs must be confined in a greater concentration than would normally be the case, allowing them access to a small plot at a time and then moving them on as soon as the ground is churned up. Again, electric fencing is the best way of controlling them. There are quite a few smallholders these days who use their pigs to do their winter digging. The pigs are fenced off in one section of vegetable garden at a time, and do a marvellous job of manuring and turning the soil in readiness for spring cultivation. Traditionally, the area ploughed by pigs was sown with an arable crop. Even with today's large-scale farmer, there are still echoes of this old practice to be found. Some cereal farmers still collect pig manure from large pig breeders and spread it on their arable fields in the winter.

Feeding

The principle of feeding pigs is to ensure that they have adequate and balanced protein, carbohydrates, minerals and vitamins for quick growth, but produce lean meat with a minimum of fat. For piglets and weaners, the ration will require approximately 18 per cent protein, while fattening pigs need about 16 per cent protein. Proprietary pig meal, which is available as powder or cubes, contains a balance of the appropriate nutrients. The cubes can be fed dry in a trough, but the powdered meal needs to be mixed with plenty of water.

A dry sow will need approximately 2.2 kg (5 lb) of dry meal a day if she is indoors, but on grass, this can be reduced to 1.3 kg (3 lb) a day. Lactating sows need about 1.3 kg (3 lb) a day plus 450 g (1 lb) for each piglet. Fattening pigs will need about 2.2 kg (5 lb) a day, but on grass this will reduce to about 1.3 kg (3 lb) a day, as long as the grass is good quality. Outdoor pigs will take longer to fatten than indoor ones, and in periods of colder weather, will require more food to keep warm.

If skimmed milk is available from your own cows or goats, this can be used to mix with dry meal, and is a particularly good source of nutrients for fattening pigs. Small 'chat' potatoes can often be bought cheaply as a stock feed from potato growers. Boiled up, these provide a cheap and useful addition to the diet. It is important not to feed too many potatoes, however, otherwise too much fat is produced in relation to lean meat. Roots such as turnips, swedes and mangolds can be chopped up and given in small quantities to supplement the diet, and a certain amount of green fodder crops such as chopped kale can be given. No waste food that might have been in contact with meat should be given as swill. Where swill feeding is envisaged, it can only take place with Ministry approval, and in licensed premises, with authorized equipment and practices.

Many farmers grow their own cereals for feeding their pigs and mix their own rations on the farm. Where this takes place, storage facilities that are vermin proof will be needed.

Pigs enjoy windfall apples, but they should not be given too many, or they will fill themselves up on them at the expense of more nutritionally concentrated foods. Apples are a particularly useful addition to the diet of outdoor, foraging pigs. It was a common belief in the past that fattening pigs needed apples in their diet to improve and enhance the flavour of the meat. When one remembers that apple sauce is an essential accompaniment to pork, there may be some truth in it.

Breeding

It is obvious that a boar is needed for breeding purposes but, as boars can be dangerous creatures, it is not worth the small farmer keeping one, unless he is particularly interested, and equipped to do so. A boar needs his own house and exercise yard, to which the sows are introduced only for servicing. Some farmers who keep pigs outdoors do run a boar with their breeding sows. People who specialize in keeping older, rarer breeds may also find it better and more convenient to keep their own boar. For the average small farmer, it is much better to utilize artificial insemination.

A gilt, or young female pig, is ready for breeding from about six to eight months of age. The signs of being 'on heat' are unmistakable.

She may try and mount other pigs, and will stand as firm as a rock if you place your hand on her back. In fact, if you try to move her at all, you will find it virtually impossible.

If a boar is kept, the gilts are often housed next to his pen, and are introduced into his quarters, one at a time for servicing. It is normal to repeat the mating after twelve hours in order to ensure that the service has taken. Some pig units use a second boar for this.

If artificial insemination is used, it is vital to spot the first signs of heat as soon as possible. Restlessness, lack of interest in food, a swelling of the vulva and a tendency to try and break out of confinement should all be watched out for, in addition to the well-known 'standing'. The AI centre should be contacted immediately signs are noticed. The local MAFF office or the local vet will advise where the nearest one is. There is no problem about administering the catheter holding the sperm, for the sow will stand still while it is inserted in the vagina. It is easy for the novice to learn the procedure, but the first time, it is as well to have an experienced handler to demonstrate.

Pregnancy lasts three months, three weeks and three days – give or take a few days, and during this time the sow will need to have adequate rations to cater for her own needs as well as those of her piglets.

If the sow is outside, she can be brought in about a week before farrowing and given a wash to ensure that she, and in particular her teats, are quite clean. It is also a good idea to treat her against mange and lice by acquiring the appropriate skin medication from the vet. No traces of this should remain on the teats, in case the piglets ingest it while suckling.

FARROWING

Farrowing (giving birth to piglets) is best carried out in the protected conditions of the farrowing house. This does not need to be a highly specialized building, as far as the small farmer is concerned. Any sound building that gives weather protection and has electricity can be adapted. The house should be cleaned and disinfected before occupation, and the floor covered with a thick layer of straw. The sow will then decide where her nest is going to be, and will arrange the straw accordingly. She will also decide which is to be her dunging area, and it is important to remove the dung every day.

One side of the farrowing house should be designated the piglets area, and an infra-red warming lamp placed above the spot. This is divided off from the sow's area by a crush bar. This is a metal bar placed in such a way as to exclude the sow from the area, but still allowing her to see the piglets. The reason for this is to protect the piglets against the possibility of being crushed by the mother when she rolls over. They go to her to suckle, but as soon as they have finished they are attracted to their little area by the warmth of the lamp. It must be said, however, that the incidence of piglets being crushed is small, and is probably more likely in large units where there is not the same degree of relationship between the stockman and his animals. I have never used a crush bar, and have never lost a piglet in this way. The sows have always demonstrated considerable care and attention to their piglets, appear to know exactly where they are, and give a continuous and contented series of grunts while suckling them.

While the sow is giving birth, it is best to be there and to place a box under a lamp, ready for the piglets. As they are born, they can be checked for deformities, and to ensure that their nasal passages are clear, given a quick rub with a towel, and placed in the box until they have all arrived. Once the birth is over, the piglets can be gently placed near the teats, where they will immediately begin to investigate and suck. The first milk is vitally important because it contains colostrum, a specially concentrated food with a high level of nutrients and antibodies from the mother, which gives protection until the piglets can produce their own.

Occasionally, more piglets are born than there are teats to feed them, and in this case it may be possible to introduce them to another newly farrowed sow. Failing this, it will be necessary to hand-rear them, like orphan lambs. They should, however, have colostrum. There is often a 'runt', which is smaller and less vigorous than the others. Again, hand-rearing

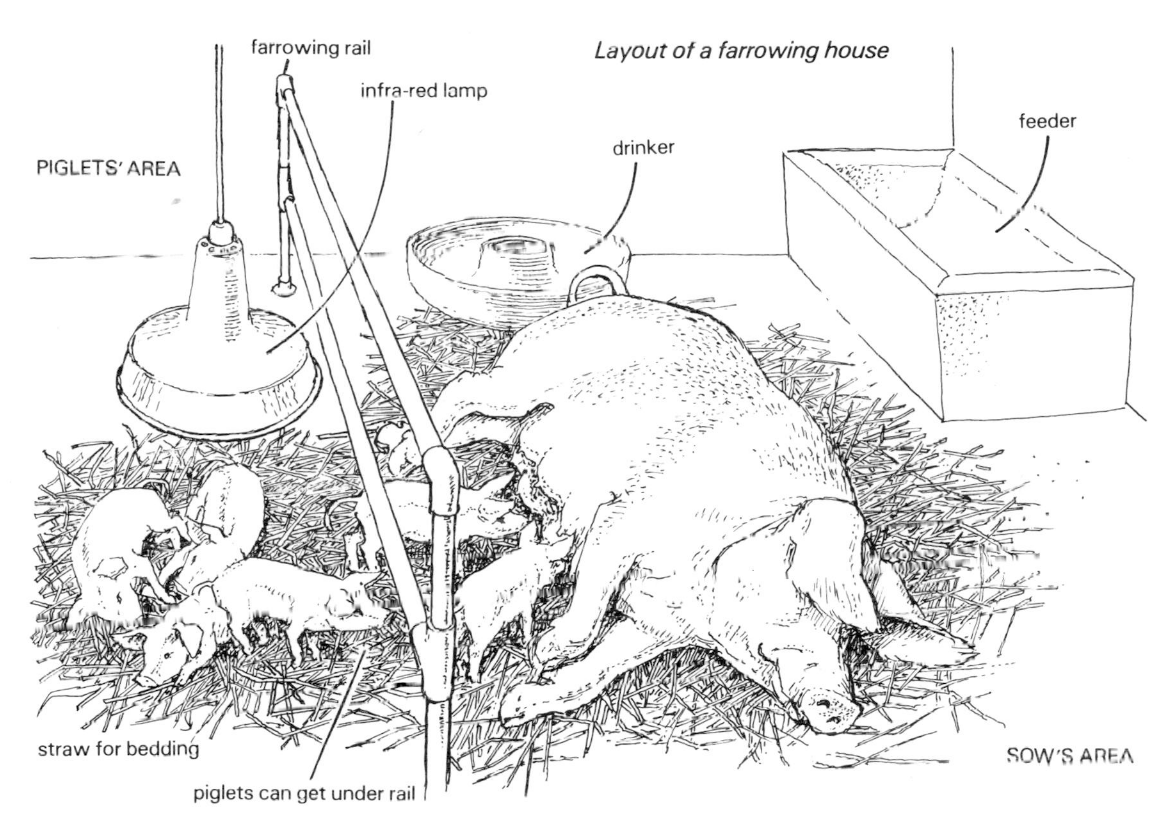

Layout of a farrowing house

may be necessary, but often this proves to be a waste of time as they usually die anyway. Such piglets, even when they do survive, never do as well as the rest of the litter, and have little commercial value.

For the first two days after farrowing, the sow should not be given more than her normal ration of concentrates, but from the third day onwards she will need approximately 1.3 kg (3 lb) of dry meal a day, plus an extra 450 g (1 lb) for every piglet she is suckling.

The piglets will be content with milk from the mother for two weeks, but by the third week, they will be ready for what is called 'creep feeding'. This is a beginner's ration and is available as a specially formulated proprietary feed from agricultural feed suppliers. Alternatively, it can be given as a mixture with some flaked maize. Water should aways be available.

WEANING

In intensive units, weaning can take place as early as three weeks, but the normal time is around eight weeks. During this time, the amount of feed taken by the piglets will have increased as they grow. The lamp providing warmth for them will have been gradually raised to provide a cooler temperature, until, when they are hardy enough, it will be removed completely. When this is will, of course, depend upon the overall temperature and on the readiness of the piglets themselves. If they are seen to be huddled together, shivering, then it is obviously too soon to remove the heat.

Around about seven weeks, a weaner meal should be introduced. Again, this can be a proprietary ration, unless you are mixing your own meal from home-grown cereals. The sow should be removed at this time, for as the piglets eat more concentrates they will be less interested in milk anyway. The young pigs can either be left in the farrowing house with the bar removed so that they have access to the whole room, or can be transferred to other roomy quarters. Once the weather is mild enough, they can go out onto fresh pasture. Weaning time is also a good time to worm them, in case they have picked up worms from the mother.

Health and general management

Avoiding trouble is largely a matter of common sense. It has been pointed out that pigs are essentially clean animals, but they do rely on their owners to keep their quarters mucked out. The pigs do as much as they can to help by dunging in one area. Feeders and drinkers should be regularly washed and farrowing houses cleaned and disinfected before and after use.

Newly born piglets are vulnerable to anaemia, and it is normal practice to administer an iron injection during the first three to four days. The injection is normally given in the muscle at the top of the back leg. Outdoor-reared pigs seldom suffer from anaemia. The four canine teeth of piglets are long and sharp, and in large pig units these are usually clipped until they are level with the gum. This is done to avoid possible damage to the sow's teats. Many small farmers find that it is not necessary. If it is deemed to be necessary, special clippers are needed, and the technique should be demonstrated by a vet or experienced pigman.

Tail docking is another controversial practice. In intensive units, the tails of piglets are docked after birth leaving a short stump, so that tail biting is avoided. In the overcrowded conditions of the factory pig farm it is not surprising that the pigs turn on each other, but where pigs are kept more naturally and humanely, the practice is unnecessary. Contented pigs rarely attack each other.

Large white sow with her piglets, on free-range.

Castration of male piglets is normally carried out a day or two after birth. There is a certain amount of controversy as to whether it is necessary, but butchers and other buyers of meat tend to avoid boar meat. It is necessary to establish with your particular buyers whether this is the case.

Worming should take place after weaning, and at regular intervals afterwards. The frequency will depend on whether the pigs are indoors or outside, and the vet will advise on this. He will also advise on suitable vermifuges.

The skin of a pig is liable to a number of conditions, particularly lice and mite infestation. Patches on the skin should be watched for, and a proprietary medication applied. Outdoor pigs should never be left without shelter, even in the summer. Hot summer sunshine can give them sunburn and in severe cases, sunstroke, although the latter is unlikely in Britain.

Where identification of the pigs is necessary, they can be marked in one of several ways, and the best time to do it is when the piglets are about one to two weeks old. Ear tags may be used, or tattoos on the inside of the ear.

There are several notifiable diseases that can affect pigs. These are foot and mouth disease, swine fever, swine vesicular disease and Tescher disease. Swine vesicular disease is difficult to distinguish from foot and mouth disease.

Veterinary advice should be sought in all cases of unexplained illness that show no signs of clearing up. There are obviously conditions, such as constipation, which occasionally arise, where a dose of pharmaceutical grade liquid paraffin works well but such conditions clear up quickly.

Breeding stock will need to have their nails clipped periodically.

Selling pigs

The Fatstock Marketing Corporation is a large organization that sells farmers' meat livestock in Britain. Unlike many buyers, it buys meat on a deadweight basis, and it owns abattoirs and packing plants in different parts of the country. The FMC will buy pork and bacon pigs. Farmers' markets operate in most market towns. Stock is sold on a liveweight basis to the meat buyers who come to the auctions. Weaners can be sold in markets, or direct to farmers who wish to buy them for further raising.

Arrangements can be made with local butchers who may be interested in finished pigs, but it is important to find out whether he also takes uncastrated boars. Weaners, older pigs and breeding stock are also sold through advertisements in the farming media. In the case of the traditional breeds, it is often more effective to advertise in the smallholding magazines, or in those catering for rare breeds interests.

Whatever the fate of the pigs, arrangements need to be made in good time before they are moved. The necessary permit needs to be obtained from the MAFF divisional office, the auctioneers need to be contacted so that the appropriate entry is made, and the transport needs to be arranged. Pigs cannot be transported with other animals, and they must also be carried in vehicles suitable for the transportation of animals. Again, the local office of the MAFF will advise on this.

All pigs must be slaughtered in an abattoir or licensed slaughterhouse if the meat is for sale. Pigs can be slaughtered on one's own premises by a licensed slaughterer, as long as the meat is for one's own private use. Slaughtering must be preceded by stunning with a purpose-made stunner.

10 SHEEP

Sheep are kept for their wool, for their meat or, in some specialized cases, for their milk. (Information on milk sheep is given in Chapter 6.) With a wide variety of breeds to choose from, there is a problem in selecting the most suitable type. Here, it is the available land that dictates which one. Although sheep will generally do well on poorer pasture, unless it is particularly wet, the breeds have evolved or been developed for different conditions. In hilly areas, the sheep are generally smaller and hardier than the lowland breeds.

If more than four sheep over the age of four months are kept, and the producer wishes to sell the wool, he cannot sell it privately but must be registered with the British Wool Marketing Board. The Board has undoubtedly improved the overall situation with the marketing of wool, and has improved standards, but those gaining the most benefit are the large producers. The small sheep producer often finds himself in the galling situation of being paid a low price for his fleeces, when he could gain a far higher price if he had been able to sell them direct to spinners and weavers, and craft shops. Faced with this situation, many small producers concentrate on selling lambs, or breeding stock if their main interest lies in the older, rarer breeds. Some get round the situation by selling fleeces from slaughtered sheep after coming to an arrangement with the meat buyer and the slaughterhouse. Sheep farmers in hill areas qualify for a subsidy or compensatory allowance. There is also a fat sheep guarantee scheme which provides for payments to be made on clean sheep or their carcasses which comply with standards of quality and weight. Details can be had at divisional offices of the MAFF.

Again, a record of movement must be kept, and in certain areas an additional permit may be required, if local conditions make it necessary.

Breeds

The best breed to keep is the one suited to the area, but it is not quite as simple as that, as there has been much cross-breeding of different types. If the aim is to produce meat lambs, a meat breed is generally the one to choose. A meat ram, such as a Southdown or Suffolk, may be mated with a general-purpose breed, such as Clun Forest, in order to combine rapid growth rate and a heavy yield of flesh in the lambs. In less favoured, hilly areas, however, it might be more appropriate to concentrate on sheep that are less productive, but hardier and more adapted to exposed conditions. The main classifications of sheep are:

Longwool breeds These are big sheep with heavy fleeces and long wool, more suited to wool production than other types. Examples are Leicester, Border Leicester, Romney Marsh and Wensleydale.

Shortwool breeds These have shorter, finer fleeces and produce good meat lambs. Examples are Wiltshire Horn, Dorset Horn, Devon Closewool and Ryeland.

Down breeds These breeds also have short wool and the rams are often crossed with upland ewes for producing meat lambs. Examples are Southdown, Hampshire, Suffolk, Oxford, Dorset Down and Shropshire.

Hill breeds The traditional mountain sheep are small and very hardy, and are good for

Above right: Leicester Longwool.

Right: Devon Closewool.

Above: Jacob ram: a breed that is popular with home spinners, but which can be difficult to confine.

Above left: Suffolk. *Left:* Swaledale

lambs. The wool is not of as good a quality as lowland breeds, but is excellent for the manufacturer of coarser wool such as that used in blankets or tweeds. Examples of hill breeds are Welsh Mountain, Scots Blackface, Cheviot, Herdwick and Swaledale.

Semi-upland breeds These are sheep that have originated from mountain breeds, but which may have been crossed with other breeds over the years. Examples are Lleyn, Clun Forest and Speckled Face.

Rare breeds These are the old, primitive breeds, which are currently enjoying something of a revival. They are generally extremely hardy and do well on poor, sparse pasture. They include Jacob, Soay, Hebridean, Manx Loghtan, Orkney and Shetland. Jacobs and Shetland are particularly popular with spinners. Those who are interested in the rarer breeds are advised to contact the Rare Breeds Survival Trust.

It should be realized that the above-mentioned classifications are only a general guide, for there has been much cross-breeding for commercial purposes. As already referred to, the best choice is one which is suited to the particular locality, and here local opinion from both sheep farmers and MAFF officers is the best guideline.

Buildings

Sheep are essentially creatures of the great outdoors, and traditionally have never had buildings. It must be said, however, that to have some kind of building for lambing makes life easier for the owner, as well as saving the lives of a good many lambs. A delicate breed such as the Fries Melkschaap milk sheep does need shelter for most of the year and details of this are given in Chapter 6. The shelter described for milk sheep is ideal for any breed of sheep brought in for lambing (see page 89).

Sheep hurdles are virtually indispensable to the sheep enterprise. They can be used to make temporary pens for all manner of reasons, and are particularly useful in the construction of individual pens within the building where sheep are brought in to lamb. They are invaluable in making handling pens for the control of sheep at times such as dipping and marking. Most agricultural equipment suppliers sell them, and they are also advertised in the farming press.

Fencing is important as some breeds, notably the mountain sheep and some of the primitive breeds such as Jacob and Soay, are notorious for their ability to escape. Where adequate permanent fencing is not available, electric fencing is the best form of control.

Feeding

The most important part of the sheep's diet is grass, and they will spend most of the year grazing. The grass will have the highest nutritional value in the spring and summer. From autumn onwards, when the grass declines and stops growing, they will need supplementary feeding. Hay is the main feed source, but concentrates in the form of proprietary pellets or cubes can also be given. Silage can be fed, and barley and oat straw are also readily accepted. Willowbark is popular with sheep as well as goats, and twigs or branches from coppiced trees are a useful source of supplementary winter feeding. They also provide much interest value to the sheep, a factor not to be undervalued in confined situations. Milk sheep will need a concentrate ration all through the year, in addition to grass and hay. This will be somewhere between 450 g–1.3 kg (1–3 lb) a day, depending upon the level of milk production.

Sheep that are to breed need special feeding before and after mating, and this is dealt with in more detail in the breeding section. Troughs for the feeding of concentrates are necessary and these should be low enough for access without allowing food to be contaminated by droppings. Hayracks are useful in the winter housing, and a mineral lick is essential, to ensure that no mineral deficiencies occur.

Breeding

BUYING BREEDING STOCK

There is more to breeding than merely turning out a ram to run with the flock. Ewes that are selected for breeding should be sound and healthy, and have a good udder. Late summer is a good time to purchase breeding stock, and it is important to buy ewes which come from a flock with a record of good lambing, rapid weight gain and good carcass grading. The Meat and Livestock Commission runs a National Sheep Recording Scheme for commercial and breeding flocks, and the information it provides is extremely valuable for those selecting new stock.

An examination of the teeth will reveal the age of the sheep. A year-old will have two incisors at the front. These will be seen to be bigger than the remaining teeth. Such a ewe will be a first-time breeder or yearling ewe, and is the best to purchase for a future breeding flock. A sheep with teeth missing is referred to as 'broken-mouthed', and is probably old and of little use in a flock.

It is important to check the feet of the sheep, to ensure that there is no foot rot or lameness. The feet need regular trimming for, like goats' feet, the nails grow round and over the soles of the feet if not checked. The method of paring the feet is the same as that detailed for goats on page 88. If a ewe has foot rot, it should not be bought, and if it is in your own flock it should have immediate attention. Again, details will be found in the section on goats. Where sheep are first introduced onto your land, it is a good idea to run them through a foot bath containing a 6 per cent solution of formalin. The latter is available at agricultural suppliers and from vets.

FLUSHING

Flushing is a system of feeding up the ewes before they meet the ram. It applies to newly bought ewes as well as ewes from an existing flock which have been grazing all summer. They are given richer pasture to bring them up to peak condition so that multiple ovulation is more likely, resulting in more lambs. A ewe that produces twin lambs is obviously more profitable than one with a single lamb.

A good way of providing pasture for flushing is to turn the ewes out onto grass that was cut for hay earlier in the year, and which has been allowed to grow on as 'foggage', or late grazing. Grass with a high percentage of cocksfoot is a suitable one for this purpose. Grazing a fodder crop grown specifically for them is another possibility, and turnips are suitable because they are quick-growing and can follow an earlier crop. Where good grass or a grazing crop is not available, the diet can be supplemented with barley, at the rate of about 6 oz per ewe per day. This is an expensive practice but, in view of the resulting increase in the number of lambs, may be worth it.

Just before the ewes are turned out onto their new pasture they should be wormed, and yearling ewes vaccinated against clostridial disease, if this has not already been done. As they are being handled, it is a good opportunity to make a quick check on their feet and on their general condition, because once they are pregnant, handling should be kept to a minimum. (Further details on handling are given later in the chapter.)

THE RAM

The choice of ram is important, for he will have a 50 per cent determination factor in the quality of lambs the following season. Where rare breeds are concerned the choice will tend to be a ram of the same breed as the ewes. Where fast-growing fat lambs are aimed for, the choice will be a good strain of one of the meat breeds. His feet will need to be in good condition, otherwise he will not be able to service all the ewes. It is important to check with the vendor that the ram is being sold as 'warranted fertile'. If you buy on this basis, then you can return him and get your money back if he fails to get the ewes in-lamb. Check his teeth to see how old he is, as well as his general condition. Both testicles should be descended.

Where a ram is already established on the farm, he should be checked over in good time before meeting the ewes. This allows any problems with his feet, or anything else, to be cleared up. Rams are normally docile out of season (although small children should always be kept away from them), but from September onwards it is best to pen them on their own, or with an old ewe for company. As they come into season they become aggressive. While they are penned, they will need hay and green food, but it is important not to overfeed them as that will make them fat and lazy.

TUPPING

Before they are turned out with the ewes, the rams should be fitted with a sire harness. This is a harness with a colouring agent, so that as a ewe is mounted during mating, a coloured mark is left on her back. It is a method of checking which ewes have been served and which have not. Different colours are used, with the colour being changed every sixteen days so that the date of lambing can be worked out. If ewes which have already been served are seen to have a second colour, this is a sign that they have come into heat again and were not pregnant the

Ram fitted with a sire-harness.

first time. If this is widespread, it is a sign that the ram, although active, is infertile, and should be replaced immediately. Other problems which may develop are the seeming infatuation of the ram for a particular sheep. It can become so marked that the ewe has to be taken out of the flock for her own protection. Where a flock is made up of different breeds there may also be a tendency for the ram to select some and not others. During the period of 'tupping' or mating, the shepherd needs to keep a careful eye on what is happening, to make sure that all the ewes become pregnant. A young ram can serve up to about thirty ewes, while an older one can manage about fifty, and the normal tupping period is about seven weeks.

Twin lambs are common.

THE PREGNANT EWES

Pregnancy last for approximately five months and during this time the ewes should be fed adequately to cater for their needs, without overfeeding and making them too fat. Once the foggage pasture has been eaten, they can be moved around harvested arable fields to graze among the stubble, and to eat other crops such as swedes, turnips and sugar beet tops. The amounts can be controlled by electric fencing.

Once the weather really worsens and winter sets in, it is a good idea to bring the ewes to a sheltered place. This may be a shelter such as that shown on page 89, or, where no building is available, a series of hurdles and straw bales against the prevailing winds. Hay, 450–900 g (1–2 lb) per day, will be needed to replace the grass, and concentrates should be given during the last two months of pregnancy so that 'steaming up', or making adequate provision for the growing lambs, takes place, starting at 225 g ($\frac{1}{2}$ lb) per day and increasing to about 675 g ($1\frac{1}{2}$ lb) per day.

LAMBING

Ewes about to lamb are restless and will move away from the rest of the flock. Where the facilities exist, it is better to separate her from the others by putting her in a pen on her own. This is obviously easier for the small flock owner, who will usually have fewer ewes lambing at any one time. While the sheep is giving birth, she adopts a characteristic position with her head pointing straight up at the sky. Unless there are obvious difficulties, there should be as little interference with the birth as possible. Occasionally a breech presentation occurs, and in situations of that kind, it is better to obtain the help of an experienced sheep handler or vet.

Sometimes one is faced with the question of what to do with orphan lambs. If another ewe has just lambed and lost one of her own, it may be possible to introduce an orphan to her, but this must be done as quickly as possible. The orphaned lamb must be rubbed with some of the afterbirth of the foster mother. This may persuade her to accept it, thinking that it smells of her own. However, lambs are very difficult to foster with other ewes, and hand-rearing is often more successful. It should be given some colostrum from one of the other newly lambed ewes, to have a reasonable chance of sturdy growth. Bottle-reared lambs are undoubtedly attractive little creatures, but the problem is that when they get older, they never really adjust to the flock, and follow their master or mistress everywhere. Automatic lamb feeders are available that work on the principle of a central tank filled with milk substitute, the lambs sucking attached teats. It is not to difficult to construct such a contraption oneself, and it can also be used for feeding goat kids. Milk substitute is available from feed suppliers, and the instructions should be followed carefully. The quantities will be similar to those recommended for goat kids in the goat section.

DOCKING AND CASTRATION

The sheep is particularly vulnerable to blowfly attack. Its wool will often conceal a heaving mass of corruption in the form of maggots eating their way into a wound in the flesh. The tail tends to become soiled with dung and in the summer this can become a breeding ground for maggots. It is normal practice, therefore, to dock the tail, leaving a short stump, when the lambs are about one day old. The way to do this is to use an elastrator to apply a rubber ring around the tail. As it tightens, the blood supply is cut off and in about a week the tail drops off. Male lambs are castrated by applying the ring around the scrotum.

Rearing lambs

The price of lambs generally drops from midsummer onwards, so the earlier in the year that the lambs are sold, the better. Some of them might be sold before weaning, others can be weaned and sold as store lambs to farmers who are buying in for fattening in order to sell in the autumn or winter when prices go up again. One of the deciding factors is the amount of grazing available. If there is not enough, then it is better to sell. The art lies in knowing what, and how many, to sell and when. Even farmers who have been rearing and selling sheep all their lives occasionally get it wrong.

Orphan lambs will need to be bottle-fed.

Lambs going to slaughter will be approximately half the weight of an adult, and are ready when there is a firm layer of flesh along the backbone. A careful watch should be kept on the ewes and lambs, to ensure that they are getting enough food. For the first two or three weeks the lambs will feed exclusively on the mother's milk, so the provision of adequate water, grass, hay and concentrates for the ewes is vital. As a general guide, a ewe with one lamb will need about 450 g (1 lb) of concentrates a day, fresh clean water at all times and as much grass or hay as she will take. Ewes with twins or triplets will need 675 g–1.3 k ($1\frac{1}{2}$–3 lb) of concentrates a day depending upon the level of lactation. After three weeks, the concentrate ration can be gradually reduced. Concentrates, in the form of sheep pencils or cubes are available from agricultural suppliers. Mangolds are useful for feeding to lactating ewes, and about 2.7–3.6 kg (6–8 lb) a day can be given, in addition to grass and hay. Ewes with more than one lamb will, however, need a higher concentrate ration. Ewes that lamb late in the season may produce too much milk as a result of the spring flush of grass. In this case, some milk may need to be stripped out of the udder every few days, to prevent mastitis or damage to the udder.

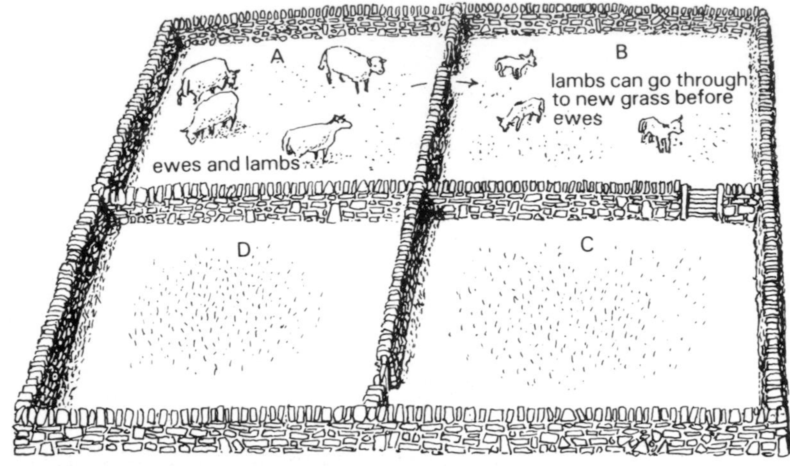

A system of rotational grazing that allows lambs access to clean grass before ewes

After a few weeks, the lambs will begin to show an interest in food other than milk, and a system of 'creep feeding' can be introduced. This is a way of giving food to the lambs in a special feeder that will not allow the ewes to gain access. In the case of early lambs which arrive before grass is available, the food will be in the form of concentrates. Again, a suitable mix is available from agricultural suppliers, and a little should be offered to the lambs, making sure that any which is left is taken away and given to the ewes, and not left until the following day. Hay can also be offered to them and they will quickly learn to nibble this.

Lambs born later when the grass is growing, should go out as early as possible with the ewes. 'Creep grazing' is often used: the new young grass is made available to the lambs first, before the ewes gain access to it. This is achieved by setting up rails or purpose-built obstacles in the field, which allow the lambs through to the new grass, but keep the ewes back. The lambs are able to go backwards and forwards, so that they can return to their dams for milk.

MEAT LAMBS

Lambs destined for sale to butchers should be weaned at ten to twelve weeks and can be sold in the market at this time, or to another farmer. Alternatively, they can be kept as store lambs. At this stage, they are frequently referred to as 'hoggets'. The fattening process aims to produce finished lambs in one of several ways. Where good grass is available, they can be turned out onto this, and given added concentrates. Root crops can be given, or barley can be utilized in the production of barley lambs. This involves fattening in housing or yards, where hay and barley are given. In the USA, alfalfa hay and maize (referred to as corn) are frequently given to housed flocks. The fattening process should take no more than eight to ten weeks, when the lambs will be ready for sale. All lambs should be wormed at six weeks old and then at intervals of four weeks but not within four weeks of sale. Ewes that produced lambs earlier in the season, should be moved onto poorer pasture so that they do not become too fat before tupping for the next season.

Sheep handling

For a large flock, a shepherd will probably use a dog to help with rounding up, particularly where sheep are grazing in unfenced mountain areas. Smaller flocks can normally be handled without such aid, but enclosures or pens of some sort are necessary. Once sheep are penned, they are accessible for whatever purpose, but they may also need to be caught and cast. This involves grasping the wool under the throat with one hand, while the other hand gets hold of the wool just to one side of the tail. Lift the back of the sheep, while simultaneously

Foot-trimming and regular checking for problems are essential tasks for the livestock-keeper.

Dipping protects sheep against a wide range of external parasites, including scab.

thrusting with your knee, and it will lose its balance, enabling it to be placed in a sitting position leaning slightly to one side. In this position its feet can be examined, and it is also ready for shearing, worming, injecting or whatever else is destined to happen. Heavily pregnant ewes should not be cast unless absolutely necessary, in which case, great care should be exercised

DIPPING

The dipping of sheep against sheep scab is compulsory in Britain. It involves submerging the animal in an officially-approved chemical dip for one minute. The best time to do this is in September, about a week before tupping. In-lamb ewes can be dipped, but many people prefer to avoid doing it at that time.

A collecting pen is required, where the sheep are marshalled before the dipping operations start, with a passageway leading from it to the dip bath. The passageway should allow one sheep through at a time, and be narrow enough to prevent turning. The approach to the dip bath should slope down into the bath itself, which will need to be about 1.2 m (4 ft) deep. It is useful to have a draining pen on the far side of the bath, because it allows the surplus liquid to drain off the sheep before they are allowed back onto their pasture. As each sheep goes into the water, it is held under the surface for one minute with a pole. The traditional shepherd's crook is ideal for this because the 'crook' can also be used to help the sheep out. Dipping in May or June, after shearing, is also necessary to control parasites, such as ticks and lice.

SHEARING

Sheep shearing normally takes place in May or June and coincides with the growth of new wool after the winter. This new wool can be seen as a clean white area underneath the dirtier, yellowish wool, and is much easier to cut than the older wool. The aim, therefore, is to part the old wool from the new and to cut off the whole fleece at the new white wool level. The sheep should be dry before shearing takes place and for this reason it is best to pen them under cover overnight. Food should also be withheld that morning, to prevent the fleeces becoming soiled by droppings. Shearing is not such a difficult procedure as some people make out. What is difficult is to do it quickly, without damaging the fleece or nicking the sheep's skin. It is advisable to watch an experienced shearer at work, and to obtain tuition at first hand. Shearing well means removing the fleece in one piece. Second cuts in an area where the fleece has already been cut will reduce its value.

Contract shearers are available to do the job for you, and they usually have their own equipment. It is up to the farmer to round up the flock and to pen them in a convenient place for the shearers.

Contractors usually use motorized shearing equipment, which enables them to shear several hundred sheep in a day. Using high-speed shearers is an expert job, as it is all too easy to cut the sheep and ruin the fleece. Anyone contemplating the use of such equipment would be well advised to gain experience by going on one of the excellent courses available on sheep handling. Both the MAFF and a number of agricultural colleges organize such courses, and these are advertised in the farming press.

If the flock is large, consider using contract shearers.

Hand-shearing is still frequently practised by owners of small flocks and although slower, is safer for the relatively inexperienced shearer. The sheep is cast, leaning slightly to the left (or to the right if the shearer is left-handed). The wool is clipped first from the face, then down the brisket or upper chest to the abdomen, so that the fleece is parted at the belly. One hand ensures that the skin is kept taut and also protects vulnerable areas such as teats and genitals. Once the wool is cleared from the belly, clip the inside legs, ensuring that each leg, in turn, is stretched out while being clipped. Place the sheep on its side and continue with the outside leg.

The wool is clipped back along the side, working from the belly to the backbone When these cuts are complete, shear from the tail upwards, keeping parallel with the backbone. Repeat on the other side, making sure that the sheep does not get up on all four feet, as it will try to bolt and will disappear over the horizon with a cape of wool billowing behind it. This was my first experience of sheep shearing, and after the escapee was caught, its fleece was retrieved in pieces from all over the hedgerows that it had pushed through. Clip the front legs next, concentrating on the inside first, then working around to the outside, and then joining up with the previous cutting by taking cuts running parallel with the backbone, up to the head. Trim the top of the head and the cheeks until the fleece is completely removed.

Release the sheep and lay the fleece on a clean surface. After removing any pieces of twig or other matter, fold the flanks inwards, then roll the whole fleece up towards the neck end, pulling out a twist of wool so that this forms a string to tie around the fleece before being tucked in under itself.

Selling wool

In Britain, anyone with four or more sheep who wishes to sell the fleeces must register with the Wool Marketing Board, who will sell the wool for him. The flock owner is not allowed to sell the fleeces himself. This also applies to wool that the owner has spun from his own sheep, as well as that which has been woven or knitted into garments. The latter point has been confirmed by the Wool Marketing Board. It is impossible for the small flock owner to make a reasonable return unless he can sell items made up from his own wool, for the Wool Marketing Board, although an excellent organization, is geared to catering for the needs of the large producer.

SPINNING

Wool is made up of fine fibres which tend to arrange themselves in bunches called 'staples'. Depending upon the breed of sheep and type of wool, these staples will be short, medium or long. Each wool fibre has microscopic scales which allow it to cling onto other fibres, giving the wool strength, and making spinning possible. During the spinning process, the staples of wool are drawn out and twisted so that the fibres, clinging to each other, form a thread or yarn. In many parts of the world, hand spindles are still used to provide the weight and momentum for twisting the fibres together while, at the same time, drawing them out. A spinning wheel is, however, much quicker, allowing a greater quantity of yarn to be produced for less effort and in a shorter time.

Spinning is currently enjoying a revival of interest and small flock owners often find that a farm 'open day' that includes traditional spinning and weaving is popular with the public. A small shop selling associated equipment, books and pamphlets can also be a good source of subsidiary income. Many courses are now offered on spinning and weaving, and anyone interested in the practical aspects is advised to go on one of these for expert tuition. Spinning is not difficult, but it takes practice to learn how to do it well. A wide range of spinning wheels is now available, including some which come as 'ready-to-assemble' packs, and which are good value.

DYEING

The hanks of spun wool can either be left in their natural form or dyed. There is a large range of chemical dyes available from craft suppliers, but natural dyes from plant materials produce colours and shades which are less harsh. The spun hanks should be gently washed to remove oil and dirt from the wool before dyeing. The principle of plant dyeing is to extract the dye from the plant by simmering in water. As a general rule, 450 g (1 lb) of plant material, such as bark, leaves or flowers, will be required for every 450 g (1 lb) of wool. The dye will be extracted fairly quickly from material such as berries, but bark will take longer. Once the water has turned a good colour, the plants are strained off, leaving the clear dye liquid.

The next step is mordanting, which is the process of creating an affinity between the material to be dyed and the dye itself. Depending upon the mordant and the plant dye, it may bring about a colour change, but will certainly make the final product more colour fast. The mordants are available from craft suppliers or large chemists and include iron in the form of ferrous sulphate, tin as stannous chloride, chrome as bichromate of potash, copper sulphate, tannic acid and oxalic acid. Some mordants are poisonous and should always be kept in sealed, labelled containers, well out of reach of children. It is also a good idea to wear rubber gloves when handling dyes. Mordanting should take place in a galvanized, aluminium, stainless steel or enamelled container. Many people find that a small boiler works well. Whatever utensil is used, it should be kept for dyeing purposes only. As a general rule, 7-14 g ($\frac{1}{4}$–$\frac{1}{2}$ oz) of mordant will be required for each 450 g (1 lb) of wool, but this is very much open to experimentation, depending upon how deep or pale a colour is required.

The mordant is dissolved in water and heated until simmering, but it is important not to boil the water otherwise the wool becomes matted. Leave the wool hanks in the mordant for about half an hour, stirring occasionally to ensure even distribution, then remove and wring gently. It can either be dried, and dyed at a later date, or dyed immediately.

To dye the wool, heat the dye water until it is simmering gently. Place the wool in it until the required colour is obtained. Then squeeze the wool hanks gently and hang them up to dry.

The following table gives a general guide to the possibilities of plant dyeing.

Plant dyeing table

The colours given are a general indication only, because so many shades and variations are possible. The mordants listed are suggestions; some plants don't necessarily need one, e.g., oak, which is high in tannin, will give brown on its own, but with alum gives gold. Most lichens and some plants, notably woad, will not easily give up their dyes unless fermented and/or boiled. The traditional method of fermentation was to leave the dye plant for several weeks or months (in some cases years) in urine. Modern equivalents are ammonia or lime. The key words here are 'experiment for yourself'. Much of the traditional knowledge of dyeing has already been lost, and is waiting to be rediscovered.

The common names given are those used in the UK. In the USA, they may be different so please refer to the Latin names if in doubt.

Common name	Latin name	Parts used	General colour guide	Suggested mordants
Agrimony	*Agrimonia eupatoria*	leaves	gold	alum, chrome
Alder	*Alnus* spp.	bark	yellow, brown, black	alum, iron, copper sulphate
Alkanet	*Anchusa tinctoria*	roots	grey	alum, cream of tartar
Apple	*Malus* spp.	bark	yellow	alum
Barberry	*Berberis* spp.	twigs	yellow	alum
Bilberry	*Vaccinium* spp.	berries	purple	alum, tin
Blackberry	*Rubus* spp.	berries, young shoots	pink, purple	alum, tin
Blackcurrant	*Ribes* spp.	berries	grey, deep purple	alum, tin
Blackwillow	*Salix nigra*	bark	red, brown	iron, chrome
Bloodroot	*Sanguinaria canandensis*	roots	red	alum, tin
Bracken	*Pteridum aquilinum*	young shoots, old tops	yellow, green	alum, chrome
Broom	*Cytisus* spp.	flowering tops	orange yellow	chrome, tin
Buckthorn	*Rhamnus cathartica*	twigs, berries	yellow brown	alum, cream of tartar, chrome tin, iron
Cherry	*Prunus* spp.	bark	pink, yellow, brown	alum
Coreopsis	*Coreopsis tinctoria*	flower heads	yellow, orange	chrome, tin
Cypress	*Cypress* spp.	cones	tan	alum, chrome
Dahlia	*Dahlia* spp.	petals	yellow, bronze	alum
Day Lily	*Hemerocallis* spp.	flowers	yellow	alum, tin, copper sulphate

Common name	Latin name	Parts used	General colour guide	Suggested mordants
Dog's Mercury	*Mercurialis perennis*	whole plant	yellow	alum
Dyer's Broom	*Genista tinctoria*	flowering tops	yellow	alum, chrome
Elder	*Sambucus nigra*	leaves, berries, bark	yellow, grey	iron, alum
Golden Rod	*Solidago* spp.	flowering tops	gold	alum, chrome iron
Heather	*Erica* spp.	tips	yellow	alum
Horsetail	*Equisetum* spp.	stalks	green	alum, copper sulphate
Hypogymnia lichen	*Hypogymnia psychodes*	whole lichen	gold, brown	
Ivy	*Hedera helix*	berries	yellow, green	alum, iron
Lady's Bedstraw	*Gallium boreale*	roots, tops	yellow red	alum, chrome, iron
Larch	*Larix* spp.	needles	brown	
Lily of the Valley	*Convallaria majalis*	leaves	gold	lime
Lombardy Poplar	*Populus nigra italica*	leaves	yellow, gold	alum, chrome
Madder	*Rubia tinctoria*	whole plant	orange, red	alum, tin
Maple	*Acer* spp.	bark	tan	chrome, copper sulphate
Mahonia	*Mahonia aquifolium*	roots, berries, whole plant	blue, brown	alum, chrome
Marigold (Try also Tagetes spp.)	*Calendula* spp.	whole plant flower heads	yellow,	alum, chrome
Meadow-sweet	*Filipendula ulmaria*	roots	yellow, green	alum, iron
Menegussia lichen	*Menegussia pertussa*	whole lichen	pink, when boiled	washing soda
Nettle	*Urtica dioica*	fresh tops	yellow, green, grey	alum, iron
Oak	*Quercus* spp.	inner bark	gold, brown	alum, chrome
Ochrolechina lichen	*Ochrolechia parella*	whole lichen	orange, red when fermented in urine, then boiled	alum
	Ochrelechia tartarea	whole lichen	red, purple when fermented in urine, then boiled	alum
Onion	*Allium cepa*	skins	yellow, orange	alum
Parmelia lichen	*Parmelia caperata*	whole lichen	yellow, brown, when boiled	oak bark
Peltigera lichen	*Peltigera canina*	whole lichen	yellow when boiled	alum
Pokeweed	*Phytolacca americana*	berries	red, tan	alum
Privet	*Ligustrum vulgare*	leaves, berries	yellow, green, red, purple	alum, chrome tin
Pyracantha	*Pyracantha angusti-folium*	bark	pink, brown, grey	alum, chrome
Ragwort	*Senecio*	flowers	deep yellow	
Silver Birch	*Betula pendula*	leaves, bark	yellow, gold	alum
Sloe (Blackthorn)	*Prunus spinosa*	sloe berries, bark	red, pink brown	alum
Snowberry	*Symphori-carpus albus*	berries	yellow	alum
St. John's Wort	*Hypericum* spp.	flower tops	red, yellow	alum, chrome
Sumach	*Rhus* spp.	berry tops leaves	tan, brown	alum, chrome
Sweet Woodruff	*Asperula odorata*	whole plants	red, pink	cream of tartar
Tansy	*Tanacetum vulgare*	flowering heads	yellow	alum
Usnea lichen	*Usnea barbuta*	whole lichen	yellow when boiled	
	Usnea lirta	whole lichen	purple when fermented in urine	
Weld (wild Mignonette)	*Reseda luetula*	whole plant	olive green	alum,
Woad	*Isatis tinctoria*	whole plant	blue	cream of tartar, lime
Yellow Flag	*Iris pseuda-conus*	root	grey, black	chrome, tin, iron
Xanthetia	*Xanthetis parictina*	whole lichen	purple, blue	

WEAVING

Weaving has also become popular in recent years, and always provides a focus of interest in a small farm or rural crafts display for the public. It is advisable to go on one of the readily available courses, and these are advertised in small farming and craft magazines. Local authorities often run night school classes in weaving.

The basic principle is that, within a square frame a set of rigid warp threads run parallel, through which weft threads are woven in and out. Within this basic pattern, there is a limitless variety of designs. There are many weaving looms available, from small table size, to larger treadle models.

SHEEPSKINS

Fleeces are taken from living sheep, but skins are the products of slaughtered animals. All animals which are destined for slaughter as meat for sale to the public must be killed in a registered abattoir, but it is possible to negotiate for the return of the skins. Lambs that are raised for one's own consumption can be legally slaughtered on one's own premises, as long as a humane killer is used and no cruelty is involved.

There is a ready market for cured sheepskins, as they are popular as hearthside or bedside rugs. There are many curing methods, but one of the easiest and most effective is to use formalin and lankroline, which are available from specialist craft suppliers and larger chemists.

The first thing to do with a skin is to scrape away as much as possible of the connective tissues, fat and flesh particles, without damaging or puncturing the skin. At this stage, the slippery membrane will not all come away, but this can be dealt with later. Immerse the skin in an old bath containing a 5 per cent solution of formalin and leave for a few days. It may be necessary to weight the skin down with a brick. Remove the skin and rinse thoroughly, then remove as much water as possible. Using a spin dryer is effective, but more delicate skins, such as those of rabbits, should be air dried. Tack the skin, hairside down, onto a board and apply a generous amount of lankroline, rubbing it in well with an old nailbrush. Now scrape away the remains of the membrane using a blade or sanding block. A sanding disc on an electric drill is effective on sheep and goat skins, but should not be attempted on rabbit skins. It is important to keep the disc away from the edges, in case it snarls up the wool. If the membrane is not all removed the first time, more lankroline can be applied and the skin left on the frame for a second session.

Once the membrane is removed, the skin is 'worked' to make it pliable. A good way of doing this is to rub it backwards and forwards over the edge of a table. Finally, any unwanted pieces are trimmed around the edges and the wool or hair is combed.

When you have perfected the technique, there is scope for curing sheep, goat and rabbit skins for neighbouring farmers. Local advertising is often all that is necessary.

Showing

All the big agricultural shows, as well as most of the smaller local events, have livestock classes where different breeds of animals are exhibited. The prize money at some shows may be worth having, but it is often the 'shop window' publicity which is of most value. Prize-winning stock is sought after, and it is easier to sell the offspring, as well as the stud services, of such animals. Most of the breeds have their own societies, and these will provide information on the type of conformation to aim for in a selective breeding programme. The older, rarer breeds are also enjoying something of a revival of interest, and there is a ready market for good stock.

Show lambs are often born earlier in the year in order to have a longer period of growth than the purely commercial stock. The wool is clipped before a show, and the coat is also 'carded' or combed with a wool carder. Faces are kept clipped so that a clear line between the face and the wool is maintained. Feet must be trimmed, as discussed for goats, and all the normal preventative measures, such as regular dipping, carried out. Purpose-made sheets are available which are tied onto the sheep, like coats, to keep them clean before a show. Leather or cotton halters are available for leading sheep, and the type used sometimes depends on the fashion set by the individual breed society. Some sheep take to being led on a halter quite naturally, but others need to be trained to it. An exhibitor needs to follow all the rules of the relevant breed society, but what most of them have in common is an insistence on the exhibitor wearing a white coat, and on the health of the animal. There are also shearing competitions at the agricultural shows, but the standard is such that only experienced professionals have a hope of winning. Most breed societies produce their own newsletter, and it is often worth advertising the sale of specialist stock in these publications.

Health

Mention has already been made of the necessity to dip against sheep scab before tupping, and against a range of parasites after shearing in early summer. Feet should be kept trimmed, and the emphasis generally kept on avoiding the onset of trouble, rather than waiting for it to appear before taking action. It is a good idea to have a footbath through which the flock runs once a month, as a precaution against footrot. This is a bacterial condition which affects sheep and, more rarely, goats, and is more prevalent on badly drained land. A 10 per cent copper sulphate or 6 per cent formalin solution will give protection and the flock should be turned out onto fresh pasture afterwards. If foot rot does affect a sheep, the first indication is lameness. An examination of the foot will reveal an evil-smelling pus oozing out between the outer horn and inner soft area. The hoof should be trimmed back, if overgrown, and the pus cleared out before applying an antiseptic or antibiotic cream or spray. In a severe case the foot may need to be bandaged for a few days.

Clostridial diseases such as pulpy kidney, enterotoxaemia and lamb dysentery are caused by a group of bacteria found in the soil. Preventive vaccines are available and these should be administered on a regular basis.

There are several diseases caused by mineral deficiencies. Hypomagnasaemia, or 'grass staggers', is brought about by a shortage of magnesium, pine is a wasting disease caused by lack of cobalt, while swayback is a nervous disorder resulting from copper deficiency. Concentrate rations include the necessary minerals, while mineral licks are also available as a supplement. The incidence of internal parasites such as roundworms, liver fluke and tapeworm can be reduced by making fresh, clean pasture available on a regular basis, but it is wise to administer vermifuges at intervals.

The sheep year

The sheep year starts in the autumn with the formation of, or addition to, the flock:

September Buy breeding sheep. Worm, and vaccinate against clostridial disease, if not already carried out. Check feet and trim if necessary. Flush breeding ewes by allowing access to good pasture before meeting ram. Check rams' feet and general condition. Dip all sheep about one week before mating.

October Tupping or mating. Allow ewes to run with ram which has been fitted with sire harness. Remember to change colour dye in harness as necessary. Give second dose of clostridial vaccine to new stock.

November December Let pregnant ewes forage arable pasture, and ensure that they are not overfed.

January Increase ewes' rations and worm five to six weeks before lambing.

February Brings ewes to sheltered area and vaccinate against clostridial diseases. Prepare lambing pens.

March Lambing. Castrate ram lambs, dock tails and earmark if necessary. Worm ewes and check feet.

April Worm lambs.

May Vaccinate lambs against clostridial disease, but not those being sold as meat lambs within four weeks. Shear sheep if weather is mild. Sell first meat lambs if weaned at ten to twelve weeks.

June Shear sheep, if not already done. Give lambs their second dose of anticlostridial vaccine. Dip sheared sheep against external parasites.

July Weaning. Keep a careful watch on ewes during drying-off period in case mastitis occurs. Give ewes a booster vaccination against clostridial disease. Worm ewes and lambs. Sell fat lambs.

August Foraging. Allow sheep to forage harvested arable fields and generally scavenge prior to flushing period. Sell surplus breeding stock at end of the month.

11 CHICKENS

During the past thirty years, chickens have been increasingly bred for one of two functions, either as egg layers for the intensive battery industry or as table birds for the broiler industry. This hybridization involved selecting and crossing the best strains of stock for the purpose, and the resulting birds are now a far cry from the older breeds, which were usually kept as dual-purpose birds for eggs and meat. The hybrid layer is light in weight and will lay up to 300 eggs a year, consuming less feed than the more traditional bird did. The broiler bird, on the other hand, is a heavy, broad-breasted bird which puts on weight rapidly without going to fat.

There are three areas of chicken-keeping where it is possible to earn a part-time income: the production of free-range eggs, the production of birds and the sale of ornamental or show breeds.

Egg production

There is undoubtedly a growing interest in free-range eggs. This is most probably the result of the associated interest in wholefoods and the various campaigns which have drawn the attention of the public to the distressing conditions of the battery units. The practice of keeping large concentrations of laying birds in battery cages so that they are unable to turn around is a slur on any civilized society, and has already been banned in some countries. A hen needs to be able to peck, scratch and take dust baths, if it is to be able to follow its natural instincts. Given free-range conditions these needs are met, and the birds will produce eggs that will have a ready market. There is no question, of course, of competing with the large producers who supply most of the shops and supermarkets. The potential market is a local one where there are no distribution problems.

Large egg producers in Europe have to comply with the Common Market (EEC) regulations by being a registered and inspected egg-packing station. The eggs must be graded and sold only according to weight. These gradings are as follows:

Size	*Weight*	
	grammes	ounces
1	70 and over	$2\frac{3}{8}$ and over
2	65–70	$2\frac{1}{4}$–$2\frac{3}{8}$
3	60–65	$2\frac{1}{8}$–$2\frac{1}{4}$
4	55–60	2–$2\frac{1}{8}$
5	50–55	$1\frac{7}{8}$–2
6	45–50	$1\frac{3}{4}$–$1\frac{7}{8}$
7	under 45	under $1\frac{3}{4}$

The small producer is exempted from these regulations, as long as he meets the following conditions: firstly all the eggs he sells are produced on his own farm, and not bought in from somewhere else; secondly, the eggs are sold to local consumers and retailers. This means that the ungraded eggs can be sold at the farm gate, in local markets, through local shops or in local hotels and restaurants. 'Local' is defined as being within the boundaries of one's own local authority and those bordering it.

In Britain, the Trades Description Act will also apply, and anyone who advertises free-range eggs when they are coming from confined birds which are not allowed access to pasture for grazing, will be contravening this act. General regulations which apply to food hygiene and quality will also apply to eggs, and the conditions in which they are laid and stored are obviously important.

In the USA, the USDA will advise on the current regulations and conditions which apply.

CHOICE OF BREEDS

If eggs are wanted, the choice will be one of the modern hybrid strains, which have been selectively bred for increased egg production. There is a snag, however. As the hybrids have been bred for environmentally controlled conditions, they may not be as hardy out of doors as the older, bigger breeds. For this reason, it is a good idea to concentrate on one of the 'brown egg' strains based on Rhode Island Red ancestry, rather than on the 'white egg' strains which are based on the White Leghorn.

One of the best hybrid strains is Warren Studler. Although bred for the intensive battery industry, it adapts well to free-range conditions, and has been found to be reliable and hardy. Different areas will have different birds available, depending upon local distributors; these are normally to be found advertising in the specialist poultry press, in local newspapers and in local Yellow Pages. The main hybrid layers are as follows: Warrens, Babcocks, Dekalbs, Hubbards, Ross Rangers, Shavers and Tetras. Individual strains of these types are normally given numbers rather than names.

Some people still prefer the older breeds for outside conditions, on the basis that they are generally hardier, although giving fewer eggs. One of the best traditional choices is the Rhode Island Red crossed with Light Sussex (RIR × LS). If available locally, it is a good choice, but it is necessary to ensure that it has come from selectively bred utility stock where the emphasis has been on breeding for egg production. It is not so much the breed, as the 'strain' which is important. One of the problems associated with the traditional breeds is that most of the best strains were bought up by the large poultry breeders for their hybridization programmes, after the last war. Most of those that remained were kept going by the poultry fanciers who were more concerned with a show appearance than with the utility aspect.

There is no doubt that a large proportion of the pure breeds now kept by small poultry keepers are vastly inferior to those utility birds kept by smallholders and poultry farmers thirty to forty years ago. There are of course exceptions, and there are still a few specialist breeders who do cater for small farmers rather than for the fanciers.

Warren Studler hen: a modern hybrid selectively bred for egg production, which readily adapts to free-range conditions.

FREE RANGING

It is important to define what is meant by free ranging. Traditionally, it meant that layers were turned out into arable fields after harvest so that they could glean the leftover seeds, or that they went onto pasture after larger grazing animals such as cattle or sheep. They were provided with a house on wheels or similar shelter, and from here, the eggs were collected daily. This

Light Sussex chickens on free range.

system usually worked well in spring and summer, although foxes were a perennial problem. The main problems arose, however, when the days became shorter and the weather became bad. Egg production gradually tapered off until it ceased entirely, and the only eggs available in the winter months were those which had been preserved in water glass or sodium silicate solution. We now know that the egg laying mechanism is controlled by hormones that are stimulated into action when light shines on the hen's eye for a certain period of time. If a bird has a minimum of fifteen hours of light a day, she will lay throughout the year. Modern free ranging can be defined as conditions that allow hens to graze pasture and that provide shelter and artificial lighting. It should be emphasized that unless the latter two conditions are provided, the hens will not lay in the winter, and the small poultry-keeper will probably make a substantial financial loss.

The pasture provided for the birds must be made available in rotation, and a piece of land that has supported poultry for one year should be rested the following year. Unless this is strictly adhered to, the quality of the grass will decline, bare patches will appear in the turf and parasites and disease will build up in the soil. This is referred to as ground becoming 'sick', and it is an apt description, for any newly introduced birds risk becoming sick as a result of infections carried over from the previous year.

MAINTENANCE OF PASTURE

It makes sense to maintain and look after pasture, for healthy grass will be reflected in healthy chickens. Overstocking should be avoided at all costs, for there is nothing that will destroy grazing more quickly than having more grazers than the land will stand. If the land is relatively light and free draining, then approximately 250 hens can be kept on one acre. Where the soil is heavy, with a tendency towards waterlogging, no more than 200 birds to the acre should be kept. After a year, all the birds should be removed from the area and the ground thoroughly raked to break up and disperse any compacted droppings. For a small area, a hand rake is sufficient for this, but for a larger plot, a chain harrow pulled behind a small tractor will do the job in a short time. Once dispersed, the droppings will soon be washed into the soil by the rain, adding to the overall fertility of the ground.

The next step is to lime the ground, not only to further improve soil fertility, but to kill off any residual parasites such as caecal or tracheal worms. The best time to apply lime is in showery weather, so that it is immediately washed in, rather than forming dust in the air above the ground. It is vitally important to wear protective clothing during this operation, including gloves, face mask and, if necessary, eye goggles. Again, a small tractor is useful for comparatively large areas, but the lime can be satisfactorily shovelled from the back of a small cart and then raked to disperse it. Agricultural lime or basic slag may be used. The former should be applied at the rate of 1000 kg (1 ton) per acre, and the latter at the rate of 500 kg ($\frac{1}{2}$ ton).

Bare patches in the turf can be re-seeded in the spring, and when not used by the birds, the ground can be temporarily grazed by sheep, goats or a cow. If preferred, the vacant plot could be used for a crop, but this would necessitate ploughing in the autumn to break up the soil, ready for drilling in the spring. If the whole plot requires re-seeding, a grazing or ley mixture which is suitable for hens is as follows:

Crested dogstail
Creeping bent
Smooth stalked meadow grass
Perennial ryegrass
White clover

These are all short-growing grasses which hens prefer to the taller, coarser varieties. The seeds should be applied at the rate of 15.8 kg (35 lb) to the acre.

HOUSING

The best type of house is one that is big enough to house the whole laying flock during periods of bad weather, and at night, and which allows them to move about and scratch in the floor litter of straw or wood shavings. Any existing outbuilding can be adapted to make such a deep litter house, and the difference between this and the traditional deep litter house, is that the birds are also allowed to graze freely on pasture, rather than merely being confined in the house. A convenient way of arranging this, and of ensuring proper rotation of pasture, is to have a house with two exits. The first opens out onto the current year's grazing plot, but the second gives access to the following year's grazing and is therefore closed off until then. When that is opened, the first exit is closed off to allow the first plot to rest. How this is put into practice will of course depend on the layout of individual farms and smallholdings.

The cost of putting up a new building is

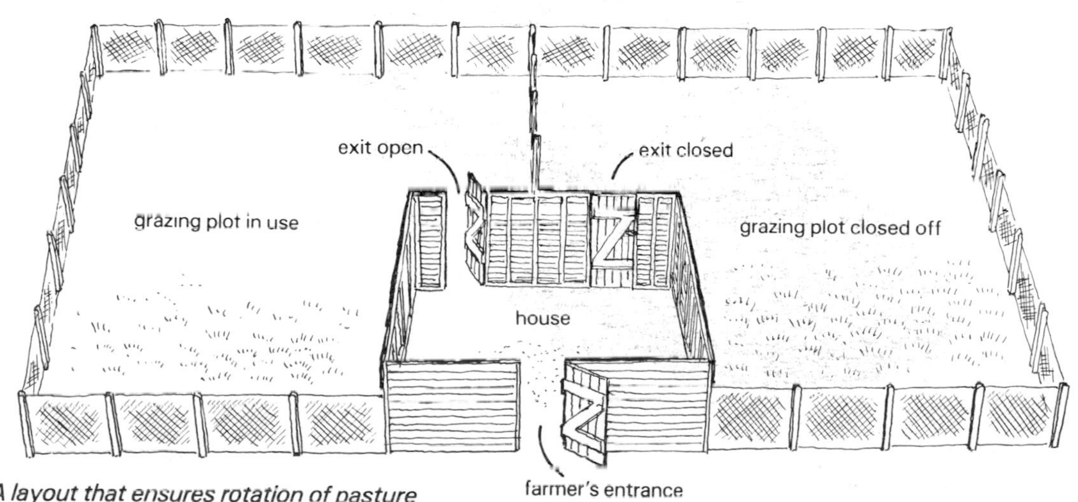

A layout that ensures rotation of pasture

A small poultry house with attached nest box and a 'pop-hole' that can be closed at night.

considerable, and for a part-time operation is unlikely to pay for itself. It is far better to concentrate on utilizing and adapting an existing farm outbuilding. Grants are not available to help with the cost of poultry units, they are generally reserved for other enterprises of full-time farmers. The agricultural world does not yet acknowledge the existence of part-time farmers, and trying to get a grant on this basis is generally doomed to failure from the start. The cost can of course be set against tax. Planning permission may be necessary for a new building and it is best to consult with the appropriate local authority.

The ideal house will have a concrete floor, for any other flooring is difficult to keep rat-proof. This should be kept covered with a dry layer of clean litter, such as wood shavings, straw or dried leaves or peat, if these are available cheaply in your area. Shredded paper has also been used to good effect. The litter absorbs droppings and can be added to as required. It is not necessary to remove the litter until that particular flock is replaced, unless the covering becomes too thick. Once every six or twelve months, the whole lot can be removed and stacked outside for composting.

Most of the droppings are passed when the hens are perching, and for this reason, many people use a droppings pit with a perch incorporated into it. In this way, the droppings will be concentrated in an area from where they are easy to remove, and the remaining floor area is kept comparatively clear.

Ventilation is important, and the best type is that which is incorporated into the roof ridge, for in this way air is freely available without causing draughts. Where this has not been installed, existing windows can be opened up and covered with wire mesh. This will allow air but exclude wild birds, who will otherwise fly in to feed. Ideally the windows should be on the south or west facing sides.

Nest boxes should be made available in the proportion of one nest to every five hens and it is a good practice to make them with sloping roofs in order to discourage the hens from perching on top of them. They will need to have clean nesting material such as hay or wood shavings, and this should be checked frequently to avoid soiling of the eggs.

Suspended feeders and drinkers will discourage the hens from scratching their litter into their food and water, and if automatic watering is available, this will be an added convenience for the poultry-keeper. Where automatic watering is installed, it may be necessary to obtain permission from the local water authority.

The hens' exit or pop-hole should be constructed in such a way that it is easily closed from the outside, so that the birds are confined for the night, thus protecting them from foxes and other predators. Hens will normally go inside without having to be rounded up and it is a simple matter to close the pop-hole once they are all inside. It is opened again in the morning.

The house should be equipped with artificial light (a strip light works well). If this has a timing mechanism in the circuit, it can be pre-set to come on and go off at the appropriate times. A hen needs fifteen hours of light in

order to lay, and as the normal daylight declines it can be extended by providing electric light. It does not really matter whether the artificial light is provided before dawn or after dusk, or a combination of both. One thing worth remembering, however, is that if the light goes out after dark, the hens will be unable to find their way to their perches. In order to compensate for this, most timing devices have a built-in dimmer which acts as a warning to the birds. Most poultry equipment suppliers stock timing mechanisms.

FEEDING AND MANAGEMENT

As a general rule, a laying bird will consume between 100–155 g ($3\frac{1}{2}$–$5\frac{1}{2}$ oz) of food a day, depending upon her size and level of egg production. She will also require rations with a 16 per cent protein value, as well as a properly balanced selection of nutrients. Proprietary layers' mash is a specially formulated ration which will provide all the necessary nutrients, in their correct proportions, and the only other requirement is water. Layers' mash is available as powder or in the form of pellets. The latter can be fed dry, as long as there is water available close by, but the powder is best mixed with water to a crumb consistency.

Laying pellets are the easiest and most convenient way of feeding, for it is merely a matter of filling up the feed container from a sack, but it is also the most expensive. This is the small poultry-keeper's biggest expense, and it therefore makes sense to find cheaper, alternative sources of feed.

If birds are free-ranging on good quality grass, they do not need to be fed exclusively on proprietary pellets, and grain can form a large percentage of their diet. Most small poultry-keepers find that a feeding practice such as the following is adequate: morning – proprietary laying pellets; afternoon – wheat; throughout the day – grass grazing.

It is important to ensure that the laying pellets are the 'free-range ration', which is normally available on request from livestock feed suppliers. This is a specially formulated 'balancer' ration that takes into account the fact that birds are receiving some of their nutrients from grass and grain. Birds which are given grain will also need to be provided with crushed oystershell so that the grain can be properly digested. This is available from feed suppliers and only a small amount is necessary. A small tin placed in the house will allow the birds to help themselves as they need it and this will probably need topping up only once every few months. Once the hens are into full lay, they may need to have extra lime in the form of limestone grit, otherwise they may suffer from a calcium deficiency. One of the problems that may arise from this deficiency is the tendency towards egg eating as the bird tries to make up for its lack of calcium by eating the egg shells. Once this arises it is difficult to break the habit, so it is best to ensure that it never occurs. Traditionally, the practice was to save egg shells and bake them in order to sterilize them, crush them up finely, and feed them back to the hens in this way.

If pellets are given in addition to grain, the ratio of one to the other will depend upon the level of grass grazing. For example, in severe weather conditions, such as heavy rain or snow, where there is little to be gained from allowing the hens to go out, the ratio of pellets will be higher than of grain – two-thirds pellets to one-third grain. In normal weather conditions, when the birds are outside, this ratio can be reversed. It should be added that birds which are allowed outside in severe weather conditions will become wet, bedraggled and possibly chilled, and a large proportion of the food they consume will go towards keeping them warm, often at the expense of egg production. It is much better to keep them confined in their house where they can move about and scratch in the litter. No one is likely to complain that the Trades Description Act is being flouted if you keep your free-ranging hens inside until the weather improves. When birds are confined temporarily, they welcome the addition of greens, such as old cabbages, which can be suspended in the house and will provide much interest for the birds.

There are some people who prefer not to give proprietary feeds at all, and who rely exclusively on grain and grass. It should be emphasized that the grass will need to be of superb quality, and the grain ration should be a mixture of different grains, rather than a reliance on one type. Wheat, oats and maize (the latter is called corn in the USA) will provide a good mixture.

EGG COLLECTION AND STORAGE

The eggs should be collected twice a day, and feeding times will be appropriate for this. A wicker or plastic basket is a suitable container for carrying them and they should be placed in a cool storage room as quickly as possible. In hot weather it may be necessary to collect the eggs more frequently than twice a day, so that they are not left in warm conditions in the hen house. Egg collection time is a suitable time for checking the state of the nesting boxes. Any soiled hay should be removed and replaced with clean nesting material, so that the eggs are kept as clean as possible. Any eggs which are lightly soiled can be gently rubbed to remove the dirt, but too much water should be avoided. Poultry equipment suppliers stock sanitizing solution which will allow egg cleaning without the risk of contamination through the shell. Any eggs which are badly soiled should, of course, be discarded rather than being offered for sale. They are quite safe to use if they are boiled for fifteen minutes.

The eggs should be stored, pointed end down, in purpose-made egg boxes or trays in a cool storage room at a temperature of approximately 10°C (50°F). The average cool pantry will meet these conditions, but if the number of eggs is large, it may be appropriate to adapt a shady outhouse as a storage room. The eggs should be sold as fresh as possible, and ideally before they are two weeks old. It may be necessary to purchase egg boxes and trays from equipment suppliers, unless there is sufficient recycling of the ones in use. Customers are usually co-operative in returning empty boxes, but obviously if the scale of operations is comparatively large, it will be necessary to buy in new ones.

SELLING EGGS

Generally there is no problem in selling free-range eggs, and friends and neighbours will probably put in regular orders. A sign outside the house will inform passing traffic that eggs are available, but this will need to be placed in such a way that it does not constitute a traffic hazard. There should also be ample parking space off the road for cars to pull in. It may be necessary to obtain permission from the local authority to put up a notice on the roadside.

Local wholefood shops and delicatessens are often interested in taking free-range eggs, but they will want regular rather than haphazard supplies, and this should be worked out at the planning stage. Local hotels and restaurants may also be prepared to take some eggs, although they are generally less interested than shops, and usually go for the cheapest supermarket eggs for use in their kitchens.

Local markets are often good places to sell free-range eggs, and the local inspector of markets can be contacted through the local authority. There may, however, be a waiting list for a market stall, so be prepared to share with someone else. This can enable you to get a stall more quickly, and means that the manning of the stall can also be shared. In some areas the Women's Institute (WI) organizes stalls where members can sell their produce.

REPLACEMENT STOCK

Commercial laying birds are replaced after one season because, in their second year, the birds eat more and produce fewer eggs. For the small egg producer, however, it is probably worth keeping his stock for two years before replacing it. It would be a far greater expense to buy in replacements every year, than to put up with fewer eggs in the second year.

Replacement stock can be bought in as day-old chicks or as point-of-lay pullets, or may be bred on the farm. If hybrids are kept, it is out of the question to try to breed replacement stock oneself, as hybridization is an extremely skilled and time-consuming occupation, and it is impossible to predict the quality of the birds unless one has access to specialist breeding stock. It is much better to buy new stock as and when necessary from a local distributor for one of the specialist breeders. If pure breeds are kept, it is possible to rear the replacements. A first cross, where two pure breeds are crossed, will produce hardy birds with characteristics of both parents. One of the best crosses, as long as the parents come from good utility stock, is the Rhode Island Red × Light Sussex (RIR × LS). This particular cross has the added advantage of being sex-linked; in other words, it is possible to tell which are males and which are females as soon as they hatch. The male chicks are yellow and the females are brown.

A consignment of day-old chicks ready to go into a brooding house until they are hardy enough for outside conditions.

Fertile eggs can either be hatched naturally by a broody hen, or artificially in an incubator. Broody hens generally make superb mothers but the snag is having them available at the time they are needed. For this reason, many people have their own incubators, so that hatching can take place as soon as the fertile eggs are available. Hens' eggs take twenty-one days to incubate and should be as fresh as possible, as fertility declines with age. They should be stored in a cool place, pointed end down, and then placed in an incubator that has been brought up to a temperature of 40°C (103°F) before the eggs are introduced. It is important to follow the instructions of your particular incubator precisely, and to remember to turn the eggs several times a day. After seven days the eggs

Artificial incubation is an easier method of rearing replacement poultry stock than reliance on a broody hen.

can be candled, or held up to a bright light in a dark room. Fertile eggs will be seen to have a radiating star-shaped pattern, indicating that the egg is fertile. Any infertile eggs should be discarded. From the twentieth day onwards, the chicks will start to break through the shells with a specially strengthened section of the beak. This is known as 'pipping'. At this time, the incubator should remain closed, and egg turning should cease. When a chick is clear of its shell, it should be left undisturbed for twenty-four hours, so that the feathers have a chance to dry off. There is no need to worry about feeding at this stage, because the chicks retain remnants of the yolk in the abdomen, which supplies them with nutrients. Twenty-four hours later, however, they will need to be removed and placed in a brooder.

Whether the day-old chicks are hatched on the farm or bought in as new stock, they will need to be protected in an area which is rat-proof and warm. They will also need to be kept separate from existing stock in order to avoid the possibility of disease transference. Any dry, rat-proof building with a concrete floor is suitable for turning into a brooding area. A section of it should be closed off with a material such as corrugated cardboard, and the floor covered with a 7.5 cm (3 in) layer of wood shavings. Paper sacks spread out on the concrete before the wood shavings are put down, make a good insulating layer. An infra-red lamp is suspended in the brooding area, and the correct height for this will be found by observing the behaviour of the chicks. If they huddle together in a mass in the centre, they are too cold and the lamp should be lowered. If they scatter to the sides, it is because they are too hot, and the lamp should be raised.

Food and water should be available for the chicks, and the drinkers should be of a type that does not allow the chicks to paddle in the water, otherwise they may become chilled and die. The best food is proprietary chick crumbs, which is a specially formulated ration with a 20 per cent protein level for quick growth. This should be available in feeders from which they can help themselves at all times.

In these protected conditions the chicks make surprisingly quick progress, and as they become more hardy the lamp can be raised progressively until it can be dispensed with entirely, when they are properly feathered. This will also depend upon the time of year and on general weather conditions. Once they are feathered, and if the weather is mild and dry, the chicks can be put out on clean pasture, in a protected run. The grass will need to be unused by previous poultry, or they may succumb to disease or parasitical infection. They may still need to be brought in at night.

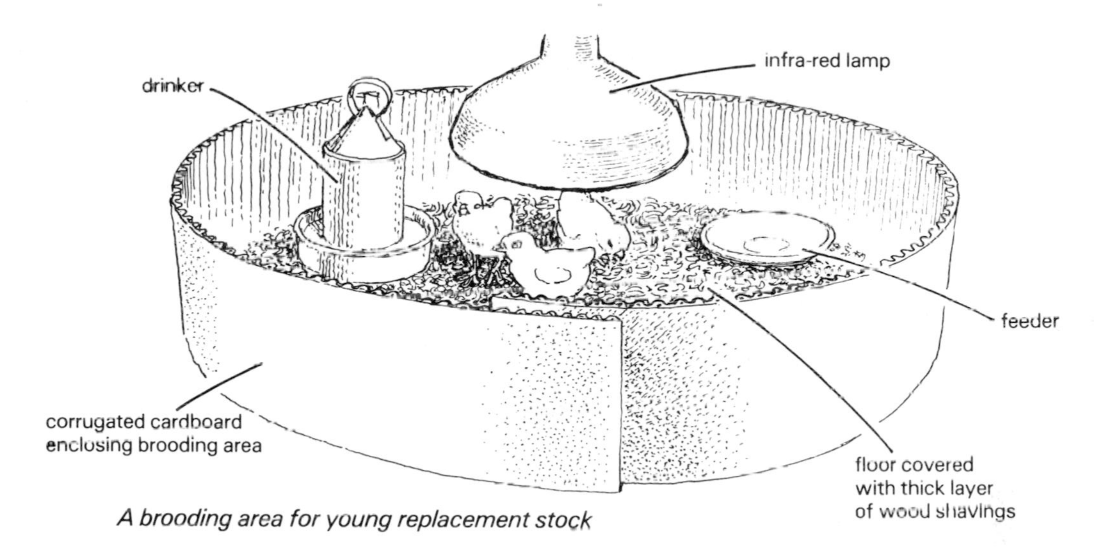

A brooding area for young replacement stock

As the chicks grow they can be switched to a growers' ration or, if preferred, they may be given a grain ration. This will need oystershell added to it so that the grains can be digested properly. A grain grinder is useful for kibbling the grain, to make it easier for the young birds to eat.

From the age of twenty weeks, the young pullets will be approaching point of lay, which is normally around twenty-one weeks. At this stage they can be given a layers' ration although, like all changes, this should be introduced gradually. Once the new stock is in full production, the old stock should be slaughtered. They can be sold off either locally as cheap boiling fowl, or to one of the specialist firms who buy batches of live birds direct from the farm (who normally advertise in the specialist poultry press). If the new birds are to be housed in the same building as the old stock, this should be thoroughly cleaned and fumigated before they are introduced, and as already emphasized they should have new, clean pasture to graze, while the old pasture is limed and rested.

Keeping chickens for the table

If chickens are to be reared specifically for the table trade, there is no point in trying to use laying birds, which are light and have been bred for egg production. It is far better to concentrate on those breeds that have been selectively bred for quick growth and for an efficient feed conversion. One of the best commercial broiler strains in my experience, is the Cobb, but any of the following are suitable: Euribrid, Hubbard, Ross or Shaver. Again, local distributors normally advertise in the poultry press as well as in local newspapers.

The broiler chicken industry is a huge business, highly developed and intensified. The emphasis is on a quick turnover and several hundred thousand birds may be kept at any one time. The birds are usually bought in as day-olds and raised in purpose-built, environmentally controlled houses. They are intensively fed until a killing time of forty-five to fifty-six days is reached, depending on the type of bird. The following are the different types produced for the commercial market:

Broiler A bird specifically bred for the table and grown to a liveweight of 1.3–2.2 kg (3–5 lb).
Poussin A bird of the same strain as broilers, but killed earlier, for the restaurant trade. The average liveweight is 450 g–1.3 kg (1–3 lb).
Capon A chemically castrated male from the same broiler strain, grown on to a heavier weight of 2.7–4 kg (6–9 lb).
Roaster A female broiler grown on to a heavier weight of 2.7–4 kg (6–9 lb).

The modern method of chemical castration used in the UK, which involves the implantation of the female hormone oestrogen under the skin of the neck of the male, is now banned in the USA and Australia on health grounds.

The small-scale producer is unlikely to be able to compete with the large producer and his market will be the local one, which is interested in home-grown, quality poultry. It is virtually impossible to find anyone selling outdoor-reared table birds because they put on weight more quickly when kept indoors. There is a growing interest, however, in birds that have been 'organically' reared and allowed to spend part of their time in outdoor conditions. The small part-time producer is in an ideal position to cater for this demand, without spending a large amount of money on housing and equipment.

THE SMALL-SCALE OPERATION

The best way of producing table birds which are humanely and organically reared, but which are not allowed to run off their weight, is to keep them in a house with an attached straw yard. This system will allow them to follow their natural instincts of pecking, scratching and taking dust baths, and will also give them access to sunlight and other outdoor conditions in a sensible way.

Any farm outbuilding will do, but a concrete floor is a considerable advantage because it will be far easier to exclude rats and mice. Where the floor is a rammed earth one, it may be a good idea to try to lay a protective layer of wire mesh and then cover it with a few inches of well-rammed earth in order to deter burrowing rats. The floor should be covered with a 7.5 cm (3 in) layer of clean, dry litter, such as wood shavings or straw, and this can be added to as necessary. There is no need to remove the litter until all the chickens are culled, when it is all cleared out and the house cleaned and fumigated before the next batch of birds arrives.

The inside of the house need only be very simple. A perch across one side will provide the birds with a place to sleep. In large, commercial units, perches are not provided. However, the natural instincts of the birds should be respected and catered for. Suspended feeders and drinkers prevent litter being scratched into the water and grain, but if wet mash is also fed, an open, heavy-based feed container will also be required. A pop-hole leads out into an outside exercise area, fenced off and with human access via a gate. This can be closed at night to confine the birds and protect them against foxes and rats. In severe weather, they are best confined to the house, so that they do not become wet and chilled, and therefore likely to lose weight and condition. If the south-facing wall is made of wire-mesh or chain-link fencing, the chickens will still have access to natural daylight.

The outside run will soon become muddy unless it is kept covered with straw, as was the traditional practice. If the layer is kept thick enough, the rain will drain through and the sun

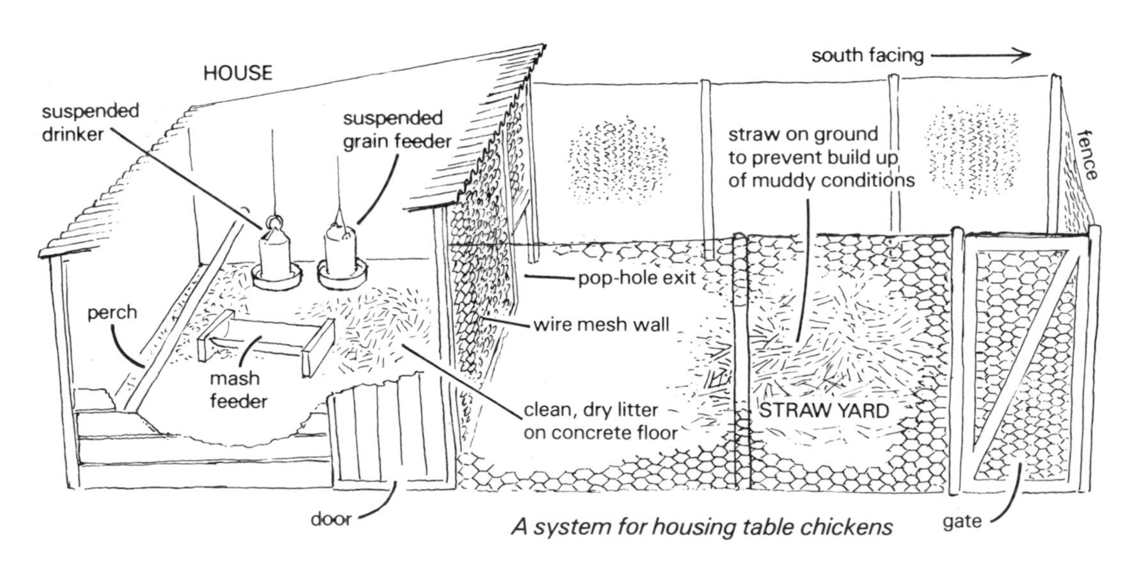

A system for housing table chickens

quickly dries the surface. As long as the birds are not let out during wet or severe weather, the straw yard will function well. It provides conditions that are close to those of the domestic bird's ancestor, the Jungle Fowl.

FEEDING TABLE CHICKENS

When the day-old broilers are first acquired, they undoubtedly have the best start in life if they are given proprietary chick crumbs or 'starter' rations, which are available from animal feed merchants. This will provide all their nutritional requirements for the first four weeks, and it should be made available on an *ad lib.* basis. They must also have access to fresh, clean water at all times. The chicks will need protection at this stage, and a brooder unit such as that described on page 135 will cater for them until they are hardy enough to do without artificial heat. This will depend on the outside temperature, but is usually by the time they are fully feathered. During the brooding stage they will need to be confined to the area of the infra-red lamp, and corrugated cardboard or some similar material can be used to make temporary walls. As the chicks grow, the walls can be removed so that they can move throughout the whole house. Once they are really hardy, they can be allowed out into the straw yard.

A typical feeding programme for commercial rearers is as follows:

chick crumbs (starter)

males:	day-old to 28 days
females:	day-old to 21 days

grower ration

males:	29–42 days
females:	22–42 days

finisher ration

males:	42 days to killing
females:	42 days to killing

These rations, in the form of ready-formulated crumbs or pellets, provide all the necessary nutrients at the appropriate times, except water. The drawback, from the small producer's point of view, is the high cost of proprietary feeds, as well as the fact that most broiler rations contain antibiotics, something that several health agencies, as well as a Royal Commission, have warned against. There is no doubt that small amounts of antibiotics in feedstuffs do contribute to a more rapid growth rate, but the subsequent meat may be regarded as containing an undesirable contaminant by the humans who eat it. If the small producer is rearing birds for the wholefood market he should make every effort to avoid these contaminants.

A traditional method of feeding table chickens was as follows:

morning:	1 part barley meal 1 part oats 1 part boiled potatoes

the whole mixed to a crumb consistency with skimmed milk

afternoon:	wheat, or a half-and-half mixture of wheat and maize (corn)
throughout the day (as available):	kitchen scraps chopped cabbage and lettuce leaves chopped nettle tops and comfrey leaves

For the small producer, this more traditional pattern of feeding has much to recommend it as it enables him to make use of a milk surplus if he has a cow or goat, as well as utilizing kitchen and garden surpluses. It is, of course, necessary to make crushed oystershell available for the proper digestion of grain. In most areas, it is possible to buy 'chat' potatoes. These are the small ones left behind after grading for the greengrocery trade, and are generally available very cheaply from growers, who sell them specifically for livestock feeding.

KILLING

Commercially, chickens are reared to fairly precise liveweights, depending on whether they are catering for the broiler, poussin, capon or roaster trade. The small producer will need to establish what weights to aim for, and this will come about as a result of local market research. It will not be necessary to be too precise; the greatest demand will be for a home-produced bird that is somewhere between 1.3–2.7 kg (3–6 lb) in weight. The small producer should, therefore, decide on a killing time that is convenient for him and take into account the amount of food consumed in relation to the amount of weight achieved. It is convenient to have a small set of poultry scales in order to check periodically on the birds' weight, and a simple way of confining a bird is to use a funnel made of canvas. The hen is placed in this while it is suspended from the scale, and the weight is established without too much fluttering.

Before killing, food should be withheld for at least twelve hours, although water should be freely available. In Britain, the Slaughter of Poultry (Humane Conditions) Regulations 1971 requires that poultry awaiting slaughter must not be subjected to any unnecessary pain or distress, must be slaughtered as soon as practicable, and meanwhile must be protected from bad weather. The Slaughter of Poultry Act 1967 requires that poultry slaughtered for purposes of preparation for sale for human consumption must be killed instantaneously, by decapitation, or by neck dislocation.

If decapitation is the method used, it is recommended that a humane poultry killer be employed. This is a simple piece of equipment, mounted on a wall, which acts as a small guillotine when the handle is pressed down. The equipment is available from most poultry suppliers. Neck dislocation is the method favoured by most small poultry-keepers, and the method is as follows. Hold the bird's legs in the left hand and, with the right hand, hold the neck just behind the head. Push down with the right hand and simultaneously twist sharply to the right. This breaks the neck immediately, but the wings will flutter because of nervous reaction for several minutes after death, and the bird should be placed on the ground while this happens so that the blood can flow towards the head. Once the fluttering has ceased, the bird should be suspended, head downwards, by tying string around the feet and hanging from a nail. It is not difficult to kill a hen, but no one should attempt it without having been shown by an experienced person.

PLUCKING

Plucking should take place while the carcass is still warm, and as soon after killing as possible, as the feathers come out more easily than if the carcass is left to cool. The order of plucking is largely a matter of personal preference, but a convenient way is to pluck the wings first, followed by the breast, back and legs. It is a good idea to wear an overall, or some kind of protective clothing, and if you have a tendency towards chest allergies it is wise to cover the nose and mouth with a face mask. If the feathers are required, they should be separated into different piles, with the big primary feathers in one group and the small, down feathers in another. It is best to pull out the feathers 'with the grain' so that the skin does not tear, as damaged skin will reduce a bird's chances of selling. The most difficult to remove are the 'pin' feathers, which are the stubs left behind on the back and legs. These can be removed with tweezers. The carcass can be singed over a methylated spirit flame to remove the last of the down, but be careful not to damage the skin.

DRAWING

Drawing is the process of removing the guts and is comparatively easy, once the knack has been acquired. A clean work surface is required, which is easily scrubbed down. A clean, muslin cloth, some sharp knives and a plastic bin for the guts are the only other essentials. The method is illustrated on the opposite page.

Unless the birds are slaughtered at a particularly young stage (poussins), they will need to have the leg tendons removed. These are strong and firm and are taken out when the feet are removed. Break the leg joint just above the foot, but do not cut through the tendons. Once these white bands are exposed they are hooked over a hook or nail on the wall, and the whole bird pulled downwards. This has the effect of drawing out the tendons. It is possible to buy a piece of equipment which will cut off the feet and remove the tendons all in one go, and this is available from poultry equipment suppliers.

SELLING TABLE CHICKENS

Large producers of table poultry are subjected to a range of regulations, but small producers are exempted as long as they fulfil the following conditions: firstly, the poultry is raised on their own farms and is not bought in from elsewhere; secondly, it is sold direct to the consumer, in local markets, or through a local butcher or shop.

This means that killed, plucked and gutted poultry can be sold at the farm gate, through shops, hotels and restaurants, in the area bounded by one's own local authority and the one bordering it. It is also possible to supply retailers further afield as long as the poultry is killed and plucked, but is not gutted and retains the head and feet. Each bird should have a label attached, with the name and address of the producer clearly marked. The latter is referred to as 'Norfolk dressed' poultry.

Health

Generally, free-range birds are hardy and do not often succumb to infectious diseases. The most frequent complaints are likely to be as a result of internal or external parasites, particularly as these can be introduced by wild birds onto otherwise clean pasture. It is important to try to prevent problems arising. The point has already been made that over-use of a piece of land must be avoided at all costs, as must the practice of allowing young stock to mix with older stock. Houses should be thoroughly cleaned and disinfected before stock is introduced, and agricultural disinfectants are available for this purpose. Feeding and drinking containers should receive frequent attention to avoid a build-up of green scum.

When buying stock, do ensure that it comes from a reputable hatchery where the breeding stock has been blood tested to ensure that blood transmitted diseases such as pullorum (*Salmonella pullorum*) are not indigenous. If you are producing your own replacement stock, ask the veterinary surgeon to do a blood test on the parent birds, to ensure that they are healthy. Day-old chicks will have been injected against Marek's disease, caused by a herpes virus, if bought from a commercial hatchery. The veterinary surgeon will provide the necessary vaccine if you are doing it yourself.

In Britain, there are two notifiable poultry

Preparing a bird for the table

1) make an incision at the back of the shank and prise up the sinews with a skewer – turn the skewer clockwise twice, pull skewer away and sinews will slip out

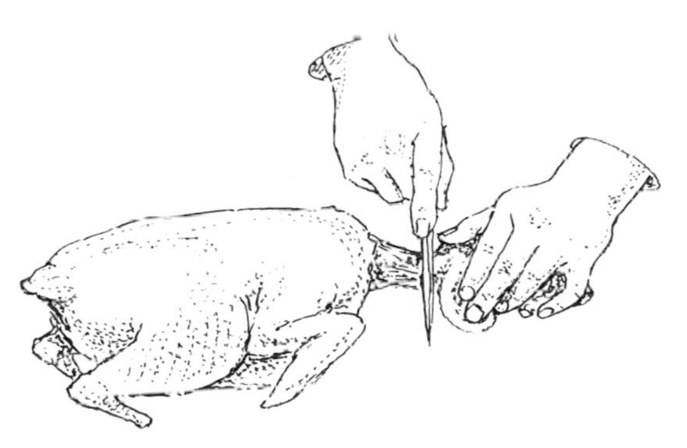

2) lay carcass breast down, cut neck skin across with sharp knife about 2·5 cm (1 in) above shoulders – slit up neck, cut through muscle at base and break the neck away – peel away crop on inner side of neck skin, and cut out close to neck cavity. Insert finger into neck cavity and loosen heart by breaking away ligaments near breast bone – prise lungs away from their attachment

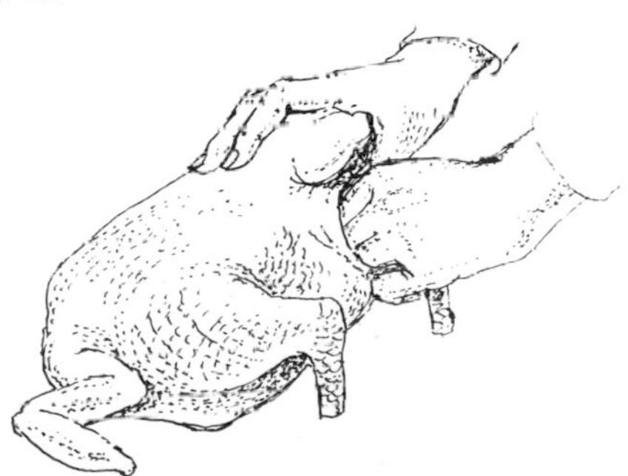

3) make a horizontal cut between the vent and parson's nose so that the finger can be inserted to hook out the intestine – cut round the vent taking care not to puncture the gut and remove with intestine attached

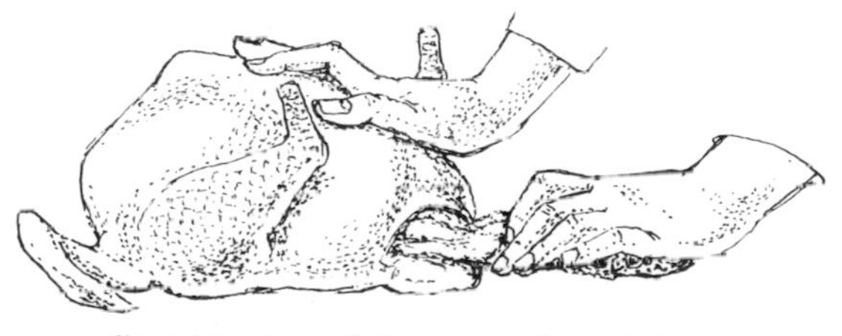

4) working through the vent cavity with fingers and thumb, free the abdominal fat and withdraw it with the gizzard – the viscera, heart and lungs can then be removed, and finally the kidneys

diseases, which must be reported to the authorities if they are suspected. These are fowl pox and Newcastle disease. If a flock is found to have fowl pox, it will be compulsorily slaughtered, although compensation is available. Newcastle disease can be treated with vaccines, and the Ministry of Agriculture will advise on preventative methods, including the correct way of vaccinating. Diagnosis is, of course, a matter for the expert, and any ailing fowl should be isolated immediately and observed for any symptoms, such as wheezing, listlessness and a drooping stance. If a single bird is affected with such symptoms (which could indicate a wide range of conditions), it is best just to leave it in isolation, to see if further symptoms develop. If it recovers, then the bird can be re-introduced to the flock. If it does not, then it is best killed. It is obviously not economic to call out the veterinary surgeon if only one bird is affected, but if several are suffering from the same symptoms, then expert advice should be sought immediately. If fowl pox or Newcastle disease is diagnosed, either the local police or the local office of the Ministry of Agriculture should be informed.

The two most frequent complaints affecting free-range birds are parasitic worms and external parasites, such as lice and mites. Rotating the use of land does keep down the incidence of parasites, but wild birds can and do re-introduce them. It is a good idea, therefore, to treat the laying birds for worms, once every six months. The veterinary surgeon will supply a poultry vermifuge, which needs to be added to the drinking water. The easiest way of dosing them is to confine the birds to their house and remove their drinkers for at least six hours. After this, introduce the drinker with the doctored water, and they will drink it all up. Normal feeding and drinking can follow, but all eggs should be discarded for the next three days. On no account should these be sold, although they are safe to give to other livestock hard boiled.

Once every three months, each laying bird should be thoroughly dusted with lice and mite powder, to kill off external parasites. Particular attention should be given to the areas around the vent and neck, and under the wings. At the same time as this is being done, check the feet and legs of each bird, in case any of them are affected by scaley leg or bumblefoot. The former is a mite infection where the burrowing action of the mite has the effect of pushing up the scales on the legs, and a white encrustation forms. The legs should be bathed in soapy water, using an old toothbrush to remove the encrustations, and the legs then painted with benzyl benzoate, obtainable from the veterinary surgeon. Bumblefoot is a hard lump on the bottom of the foot, resulting from a cut that becomes infected and then seals over. In this case, it is necessary to cut open the lump with a sharp, sterile blade and squeeze out the hard pus, remembering to apply disinfectant to the wound.

It is not necessary to treat birds that are being raised for the table, either for worms or external parasites, for they are being kept for a comparatively short time and it is highly unlikely that they will become infected.

Where litter of any kind is used, it is important to ensure that it does not become musty. This is more likely with hay than with anything else, particularly if it has been allowed to become damp during storage. Such conditions can give rise to aspergillosis, a fungus infection affecting the lungs and causing wheeziness and other respiratory problems. It can affect humans as well as birds, as the alternative name, farmer's lung, illustrates.

There is a whole range of other diseases and infections that can affect chickens, but if the birds are being kept to provide a part-time income, it is not economically viable to call in the veterinary surgeon, or attempt to treat individual birds. It is far better to isolate any ailing bird, as previously described, and just keep it under observation. If it is a minor condition, it will recover of its own accord; if more serious, the bird should be killed. There is really no point in trying to cure a bird, which even if it does survive a serious condition, will never be highly productive.

Any diseased birds which have been killed should be incinerated, not buried, to minimize the posibility of disease transference.

12 DUCKS

If you are attracted by the idea of keeping ducks there is little money to be made out of selling eggs. However, this is a generalization, and there are exceptions. This is where your own market research is so valuable. In some local markets, duck eggs can fetch more than hens' eggs and they never fail to go. Generally, however, there is still a hangover in attitudes about duck eggs, dating back to the 1950s, when they were accused of being a source of salmonella food poisoning. While this was true, they were a long way down the list by comparison with cooked and processed meats, which are still the prime source. Nevertheless, laying ducks should be kept away from possible sources of infection, such as stagnant ponds, sewage outlets or any area where decaying material may be harboured. They will also need to be confined in clean laying houses until they have laid their eggs, to ensure that the eggs are laid inside. Their tendency, otherwise, is to lay them anywhere, and to leave them at the mercy of magpies and crows.

The two other areas where a small, commercial enterprise is possible with ducks, is in the rearing of table ducklings and in the breeding of ornamental varieties. Whichever, if any, of these aspects is ventured upon, it is necessary to adopt the right system of management. What is suitable for an egg-laying strain, for example, is not necessarily suitable for a table duckling. The choice of breeds may also be crucial.

Egg production

If local conditions indicate sufficient interest in duck eggs, you may wish to acquire some layers. For 'eating' eggs, there is really only one breed to consider, and that is the Khaki Campbell. This is a brown and fawn duck with a good record of egg laying. It was bred at the beginning of the twentieth century, and incorporated genetic material from the Mallard, Rouen and Indian Runner. The male has a bronze head with a dark brown back, shading to lighter brown underneath. The wings have a bar with a greenish sheen, the bill is olive-brown and the legs are orange-brown. The female is a more uniform brown, with a lighter head than the drake, and the bill and legs are both olive-brown.

It is important to obtain stock from a breeder of commercial strains, rather than from one who concentrates on show breeds, for the latter may be more concerned with appearances than performance. It is now possible to buy Khaki Campbell strains that can lay up to 340 eggs a year, and breeders can normally be found advertising in the poultry and farming press.

HOUSING AND MANAGEMENT OF LAYERS

The important thing to remember about laying ducks is that, if they have access to water it must be fresh and clean; no build-up of litter should be allowed in or around it. Ducks will do well on pasture, but no more than 100 birds should be put on 0.4 hectare (1 acre) of ground. This is worked out on the basis of continuous occupation for two years which is the optimum period for egg production before yields begin to decline. After this it is better to dispose of the birds and acquire new stock, which will then go onto new ground. The cheapest way of acquiring replacement stock is to raise your own ducklings. An alternative is to have 100 ducks on 0.2 hectare (½ acre), but for one year only, and then to transfer them onto a new 0.2 hectare (½ acre) for the second year, while the first half is limed, rested and re-seeded if necessary. Whatever the size of the grassed area, it will need to be properly fenced to keep out foxes and other predators. 2 m (7 ft) high wire mesh fence is needed, but the cost of erecting this along such a large perimeter is prohibitive, and even if it were possible, there have been cases where determined foxes have scaled even 2 m (7 ft) high fences. From the economic point of view, it is better to have a smaller fence and invest in one of the various anti-fox warning devices which are available.

Houses are necessary to provide shelter for the ducks, particularly from high winds, which they dislike, and to provide a place for them to lay their eggs. They need only be simple structures, but should be easily moved and readily cleaned, allow access for egg collection and have a door for confining the ducks at night. The roof, and possibly the sides in particularly exposed areas, should be covered with waterproof roofing felt and there should be an overhang of several inches to allow water to run off. Ventilation is important; a meshed window or door should be available in each shelter to cater for this. Nest boxes in the ratio of one to every four ducks should be provided, with clean hay in each one. Ducks are sometimes reluctant to use nest boxes, and have a tendency to lay their eggs all over the place, but hopefully most of them will learn to use the nests, particularly if an imitation china egg is placed there initially, to encourage them.

The best type of flooring in the house is a matter of dispute amongst duck-keepers. Some prefer wire flooring, on the basis that droppings fall straight through and the houses are therefore easier to keep clean. Others prefer board floors covered with litter such as wood shavings, which will absorb the droppings. A compromise between the two is to situate the feeder and drinker in each house in a droppings area made of wire mesh, leaving the rest of the floor covered with wood shavings. This area, and the entrance to the house, should have wide ramps leading up to them, so that the ducks have easy access. They are prone to leg damage if they strain themselves trying to scramble up or down.

Khaki Campbell ducks and drake: the best of the egg-laying breeds.

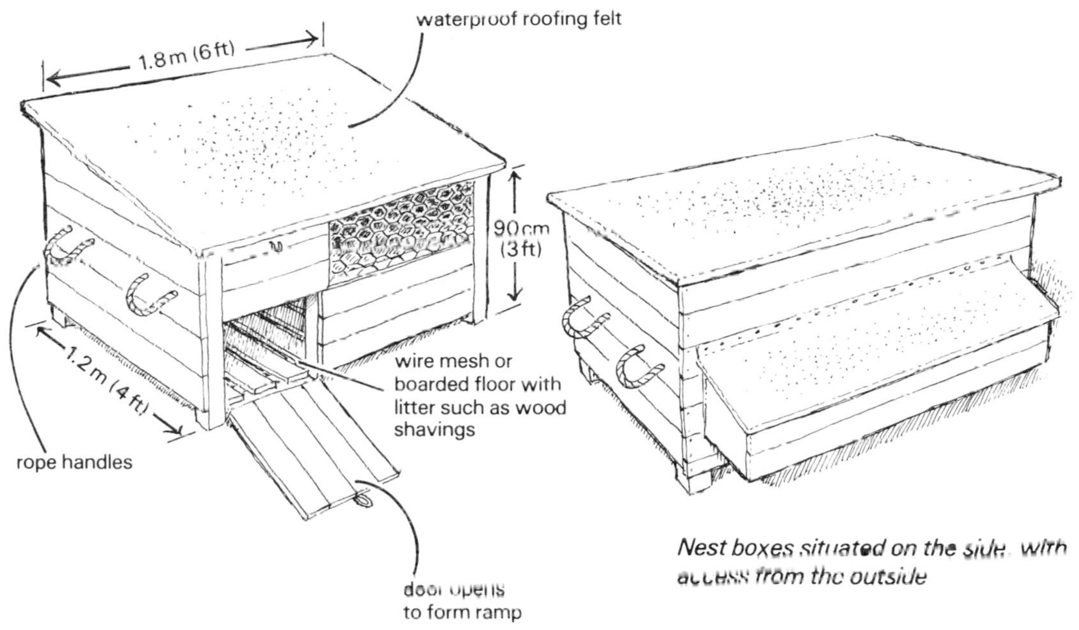

A simple, moveable shelter which provides night-time shelter for twelve to fifteen laying ducks

Nest boxes situated on the side, with access from the outside

FEEDING LAYING DUCKS

Ducks will lay well and produce good-sized eggs if they are fed on proprietary layers' rations such as that given to laying hens. The best type is the free-range mixture, which is balanced to take into account the other foods obtained from ranging activities. An adult Khaki Campbell will eat about 225 g (8 oz) of food a day. If wheat is given as well as layers' ration, this can be in the following ratio:

85 g (3 oz) layers' ration (given in the morning)
85 g (3 oz) wheat (given in the evening)
55 g (2 oz) foraged food from free-ranging throughout the day

The wheat given in the evening is best fed in the feeders in the houses so that the ducks go in without any trouble and are thus easily locked up for the night. The layers' ration may be given either in the form of pellets or as mash, mixed with a little water to a crumb consistency. Dry pellets are taken readily by most domestic ducks, but a permanent source of drinking water close by is essential.

THE EGGS

The eggs should be collected twice a day, at feeding times, and any badly soiled ones discarded; they can be hard-boiled and fed to other livestock. Any ones which are to be offered for sale should be treated in the same way as that detailed in the chapter on chickens. It is worth stressing again that duck eggs should be laid in scrupulously clean conditions to avoid any possibility of salmonella contamination. The pores in the shell of a duck egg are bigger than those of a hen's egg, and therefore absorb things more readily. Eggs offered for sale should be fresh, and certainly not more than a week old.

Table ducklings

If table ducklings are required, the most important commercial strains are the Aylesbury and Pekin hybrids. The Aylesbury has always been the table bird in Britain, while the Pekin is prominent in both the USA and Australia. In recent years, the hybridization of new commercial strains has increasingly involved crossing the breeds, with the emphasis currently on the Pekin strains with their tendency towards quicker growth. Commercial strains are advertised in the poultry and farming press. In Britain, the Duck Producers' Association represents the interests of commercial duck producers, and its advisory bureau provides information to help promote the sales of table duckling. Anyone thinking of embarking on a table duckling project should consult this organization first.

HOUSING AND MANAGEMENT

From day-old to slaughtering is six to seven weeks for commercial strains. The most usual way of housing the ducklings during this period is in a house, on litter. Some producers have their stock on wire or slatted areas to cope with the droppings, but this is not advisable in view of the ducks' delicate, webbed feet.

Any outbuilding is suitable as long as it is rat-proof and has adequate ventilation. The best form of litter is wood shavings and this should be applied to a depth of 7.5 cm (3 in) on the floor. Feeders and drinkers can be on raised, slatted or wire-meshed areas in order to confine the droppings to one place as far as possible. Details of this area are given on page 142.

Where the ground is particularly sandy, the ducklings can be penned outdoors, or even allowed to clear up the stable after harvest, as is still occasionally done in parts of Norfolk. Small portable shelters can be made available for them, but a careful watch needs to be kept for foxes. In this situation, they should not be let out until they are at least three weeks old, and they will take longer to reach killing weight. A system such as that detailed for table chickens, where a house and attached straw yard is used, is ideal.

FEEDING TABLE DUCKLINGS

Ducklings grow quickly and are normally ready for killing at six to seven weeks, having achieved a liveweight of 2.9–3.4 kg ($6\frac{1}{2}$–$7\frac{1}{2}$ lb). The growth-promoter chemicals that are normally used in broiler rations are not included in duck rations, because they pass through the digestive system before they can take action, and are therefore deemed uneconomic. This is good news for those who find the rather indiscriminate addition of antibiotics into animal feedstuffs disturbing. The normal feeding practice is as follows:

1–14 days	chick crumbs
14–21 days	duck starter ration
22–35 days	duck grower ration
36 days – killing	duck finisher ration

The rations are fed on an *ad lib.* basis, where the birds can help themselves at any time. Clean, fresh water is available round the clock. If preferred, a more traditional fattening diet can be given. One example of this is as follows:

1 part cooked potatoes or stale bread
1 part oats or maize (corn)
1 part barley
1 part wheat

If confined ducklings are given grain, they will also need a small amount of crushed oystershell so that proper digestion of the grain can take place. If free-ranging, they would normally pick up small stones themselves for this purpose.

KILLING

Killing time is critical and should be just before the ducks commence to moult for the first time. If this time is missed, it becomes much more expensive to feed them because they quickly lose condition during the moult. The stubs of the pin feathers will also be difficult to remove.

Ducks can be killed in the same way as chickens (see page 137–8), but they are more difficult to pluck, and waxing is frequently carried out. The wings are plucked first and the breast roughly plucked, then the whole carcass is dipped in wax at a temperature of 60°C (140°F). It is transferred to cold water until the wax has set, and then the wax is pulled off, bringing the feathers with it. The carcass is gutted and dressed in the same way as that of a chicken (see page 139).

Pekin ducks: the fastest growing of the table breeds.

USING THE FEATHERS

Duck down feathers, and those of geese, are a valuable commodity to manufacturers of continental duvets. If you produce a sufficient quantity at a time, it may be worth contacting the manufacturers of these, as well as those who make pillows. For smaller amounts, an advertisement in the local papers will often bring a response from local craftspeople, particularly those involved in the making of soft toys. Local branches of the Women's Institutes should also be informed, because many of their members are involved in such crafts. Craft shops are also worth approaching; they may themselves be interested in buying the feathers, and if not, should certainly be able to put you in touch with people likely to be interested. They will usually allow you to put up a card in the window, advertising your service. There is always the possibility of using the feathers yourself, of course. Attractive cushions can be made from the duck and goose feathers, and remnants of material suitable for use are available at reduced prices from textile shops.

Waxing the carcass to remove feathers is best avoided, as is wet plucking, if the feathers are to be used. Dry plucking is best, but it will be a very 'fly-away' task. It is a good idea to wear an overall, cover the head and wear a face mask. The best place is an outhouse, away from gusts of wind and draughts, and where a bit of mess will not matter too much.

Keep the different types of feathers separate – the long primary or quill feathers in one pile, the secondary, shorter ones in another, and the small, fluffy ones from the breast in yet another. The primary feathers may be of interest to those who make quill pens, or to toy manufacturers who make Red Indian head-dresses. The down feathers are the ones most sought after by the duvet and pillow manufacturers, although they may also want the tips of some of the secondary feathers; this is something that varies from one manufacturer to another.

Sometimes they will take the feathers as they are, unwashed, provided they have been sorted. Other companies may want them to have been washed and dried; the latter is certainly true of local craftspeople. It is best to avoid detergents and to use a gentler cleaner such as pure soap flakes. Make up a solution in warm water and pour in the feathers, then gently agitate the water with the hand until any dirt comes out. Rinse thoroughly, then place the feathers in old pillow cases – not more than half full – and suspend on a line in a warm place. Every now and then the pillows should be shaken up so that air circulates through the feathers and speeds up the drying process. For most purposes, where down feathers are used in such items as home-made cushions, the feathers are clean enough and will not need washing, but they should still be aired in a warm place in order to remove the characteristic 'duck' smell.

Ornamental breeds of ducks

There is a consistent degree of interest in ornamental breeds of waterfowl, and breeding pairs or trios of certain varieties can fetch a considerable amount of money. It is essential, however, to start off with excellent stock and to have researched the field thoroughly. The British Waterfowl Association is the body which oversees and protects the interests of waterfowl in Britain, and the Poultry Club also includes some breeds of ducks in its activities. In America, the organization to contact is the American Poultry Association, and there is also an International Waterfowl Breeders' Association based in the United States. It is certainly worth joining these various organizations.

It should be said however, that this type of venture is really only possible for those with a relatively large amount of land including stretches of water. While many breeds that are related to the domestic varieties, such as Call ducks, Rouens, Runners and Aylesburys, can be kept in the way detailed for laying ducks, there are many ornamental breeds which do not do well in confined situations and require semi-wild conditions. This is partly because they need a certain proportion of their food to be caught in the form of insects and larvae in the water, and also because they do not breed well in captivity. Examples are Carolinas and Mandarins.

A Carolina drake: this breed is a popular choice with those interested in ornamental waterfowl.

If the right conditions are available, the choice is between specializing in one breed or keeping several. If the latter is the case, it may be necessary to provide separate areas to prevent fighting or inter-breeding. Where a river is available, a suitable way of doing this is to fence off sections of the river, and its immediate area, so that individual breeds have their own territory, but the water flows through each section and semi-wild conditions are maintained. Ideally, the house or shelter should be placed on the highest point, with the ground sloping down towards the river. This will ensure adequate drainage.

A house similar to those provided for domestic laying ducks is suitable, although some breeds, such as the Carolinas, prefer more natural shelters, such as an old barrel strategically placed in a tree trunk. Australian Wood duck, for example, are naturally more inclined to perching in trees anyway, and this type of shelter is particularly suitable for them. It is a good idea to visit wildlife parks and waterfowl sanctuaries in order to get some idea of the range and type of shelters used.

Young stock

Ducks are not, generally speaking, good mothers, although there are, of course, many exceptions, such as the Cayuga and the Muscovy. The ornamental, semi-wild varieties are usually best left to themselves, although there may occasionally be a case for using a broody hen or an incubator to hatch out the difficult or rare ones. Where a hen is used, it is important to dampen the eggs occasionally to ensure that they do not become too dry. Duck eggs of most domestic breeds will take about twenty-eight days to incubate. The process is the same as that detailed for chickens.

The brooding of newly hatched ducklings

will, of course, depend upon the breed and the purpose for which they are being kept. Details have already been given for table ducklings, which are best kept inside. The young of a laying breed will either be in the care of their natural mother, if she has hatched them, or in that of a broody hen if one was used. In this situation, the mother (or foster-mother) and her brood are best placed in a small, protected house and attached run, such as that used for young chickens. This will ensure that they are protected against a range of predators, including rats, cats, weasels, magpies and crows.

Young replacement laying stock will go out onto new, clean pasture from the age of about six weeks, depending on weather conditions.

As far as feeding is concerned, ducklings are best fed on proprietary chick crumbs for the first six weeks. This will ensure that they are receiving adequate levels of protein and other nutrients at the crucial time of primary growth. From this point on, it will depend upon the system they are to follow. Table ducklings, as previously referred to, will follow a diet suitable for fatteners, while those destined for laying will follow the appropriate pattern already detailed. Ornamental varieties will obtain a certain proportion of their food as insects and insect larvae, from land and water. In addition to this, wheat may be given once a day, particularly in periods of greater need, such as in winter. Where the weather is very cold, a proportion of oats can be mixed with the wheat, in a 1:1 proportion, and this will give added protection against the cold.

HOW TO SEX DUCKLINGS

Once the full feathering appears, it is a simple matter to distinguish between male and female ducks, but it is not always straightforward where day-old and young ducklings are concerned. Commercially, the so-called Japanese method is used, which involves an examination of the vent. It is not an easy procedure, and is not recommended for the novice because of the danger of causing permanent physical damage to the young bird. Hold the duckling gently on its back, with the index finger and thumb of the hand holding the bird, placed on either side of the vent so that it is held slightly taut (take care not to overdo it). Now, with the other hand, gently part the vent so that it is extended and opened slightly. In male ducklings a small penis will be seen at the top, inner edge of the vent, while this is absent in females. It is best to get someone experienced to show you the technique first, before trying it out. This method may also be used for goslings.

13 GEESE

'No grass, no geese': this old adage sums up the situation precisely. They are prolific grazers and derive most of their nutritional requirements from grass. Unless there is sufficient pasture to cater for their needs there is no point in keeping them. Nor is there any point in rearing them if there is insufficient local demand for table birds.

On a permanent basis, that is, where there is no rotation of land, nine or ten heavy breeding geese will need one acre. Where young, growing geese are grazed temporarily from April to September, the stocking rate can be as high as 100 goslings, as long as the ground is not heavy or waterlogged. As soon as the grass begins to decline in September, they must be moved to another area, where Christmas fattening can take place.

If the decision is taken to keep geese and to obtain a partial income from the enterprise, there are several options: selling the goslings as day-olds or as three-week-olds, or rearing them to table weight for the Christmas trade. There is also the possibility of concentrating on ornamental breeds and selling either the young progeny or breeding trios of adult geese. This is a highly specialized area and anyone interested in pursuing this is recommended to get in touch with the British Waterfowl Association which looks after the interests of waterfowl generally, and which brings together breeders and those with an active interest in the various breeds. In the USA, the appropriate organization is the American Poultry Association.

Breeds for the table

If table geese are required, the best breeds are strains of Toulouse or Embden. These are both heavy breeds which have traditionally been developed for the table. There are, of course, many geese whose ancestry is mixed, and the average British farmyard goose is in this category. Often these are extremely good table birds and should not be discounted because no-one is quite sure what their antecedents are.

The traditional practice, in addition to the Christmas trade, was to rear 'green' geese for Michaelmas, at the end of September. This coincided exactly with the decline of pasture as the grass stopped growing, and there was little required in the way of supplementary feeding. Unfortunately, the Michaelmas festival has virtually disappeared.

BREEDING STOCK

The best time to acquire breeding stock is in the autumn, and the ideal age to commence breeding is two years old. It is in the autumn that the breeding 'sets' are formed; these are the mating groups of ganders and geese. With a heavy, table breed this ratio will be one gander to two or three geese, and the choice of partners is definitely that of the gander, not his owner. There have been many instances where a gander will have nothing to do with a particular goose, even driving her away if she comes near him. The best practice to follow is to put all the birds together in their over-wintering quarters, such as an orchard, and let them sort out their own breeding sets before the egg-laying season starts. If breeding pens are then made available, with access from the orchard, the females will lay in these, while still maintaining contact with their own gander. If, as occasionally happens, there is a lone goose left which is ostracized by the ganders, it is better to dispose of her, either by selling or eating, for she will only create a nuisance within the rigid goose regime.

An orchard is a particularly suitable place for

The Chinese geese, on the left, lay more eggs than the Embdens, but the latter are better for the table.

over-wintering of stock because windfalls from the trees, particularly the late varieties, provide a useful source of supplementary feed. Beware, however, of introducing them where young fruit trees are newly planted as geese have been known to cause considerable damage by pulling down and breaking vulnerable branches, in an effort to reach apples and pears. You should fence off the orchard into two areas, so that one half of the ground is 'resting' at any given time. This rotation of the land prevents over-grazing of the grass, which in turn leads to build-up of parasites, such as gizzard worm.

PLANNING THE LAYOUT

It is vital to plan the layout of the land before introducing the stock, and the diagram opposite indicates one satisfactory method of keeping geese without overgrazing.

To clarify the diagram: the adult breeding stock which have been introduced to the first half of the orchard in the autumn, sort out their breeding sets and, from February onwards, the geese begin to lay their eggs in the breeding pens which are accessible from the orchard. From April onwards, the young geese are introduced to the first grazing area, provided, of course, that weather conditions are suitable. Here they are left to graze until the end of September when the grass declines, at which point they are transferred to the fattening area, where they will receive fattening rations before being slaughtered for Christmas. The breeding stock has, meanwhile, been left behind in the first half of the orchard and stay there until the autumn, when they are moved to the second half.

The following February, the breeding geese are allowed access to the Year 2 breeding pens, and the young goslings subsequently introduced to the second grazing area while the first one is rested. The rotation then continues. There are, of course, many variations on this plan, depending on individual situations.

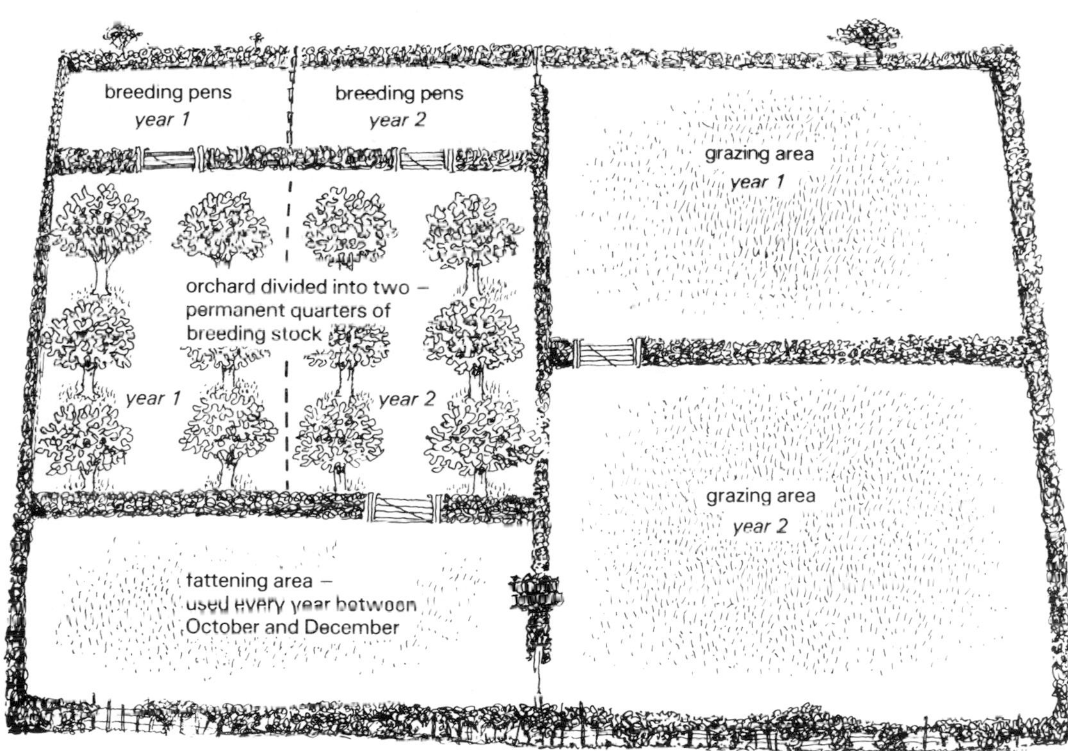

A convenient layout showing rotation of land

MAINTENANCE OF PASTURE

After the stock has been transferred, the ground should be raked to disperse droppings and limed at the rate of 1 ton agricultural lime to each acre of ground. This should be applied in showery weather so that it is washed in before it is blown away, and eye goggles and a face mask, together with other protective clothing, should be worn as a protection. If basic slag is applied as an alternative to agricultural lime, it can be applied at the rate of $\frac{3}{4}$ ton to the acre.

During the season when the grass is not being used by geese, it can be used to support other livestock such as sheep, goats or a cow, as long as no poultry are introduced. Alternatively, the grass can be cut for hay.

HOUSING

Housing for geese can be simple; they are extremely hardy, but do appreciate shelter from high winds, which are disliked by all waterfowl. Often, adult geese that are in an orchard will prefer to sleep in the shelter of trees, and ignore any housing which has been provided for them. In this situation they are at risk from foxes and it is important to ensure that protective fencing, at least 1.8 m (6 ft) high is used around the perimeters. Even this is no guarantee of safety, as foxes have been known to scale such fences.

A simple wooden structure with three sides is normally all that is necessary, and the overall length of this will depend on the number of birds using it. It should be placed facing south so that it is free of draughts and cold winds. Ideally, the structure should be easily movable and rope handles on either side will facilitate this. Naturally, the length should not be such that it is impossible to move the structure, and a bank of shorter shelters may be more appropriate.

The breeding pens will, however, need more protection because the newly-hatched goslings are particularly vulnerable to attack from rats. It is a fallacy that ganders will drive off all predators; rats have been known to take newly hatched goslings from under a sitting goose, with no defence from either the goose or gander. The shelter will need close-mesh wire floors with straw as a covering and nest material

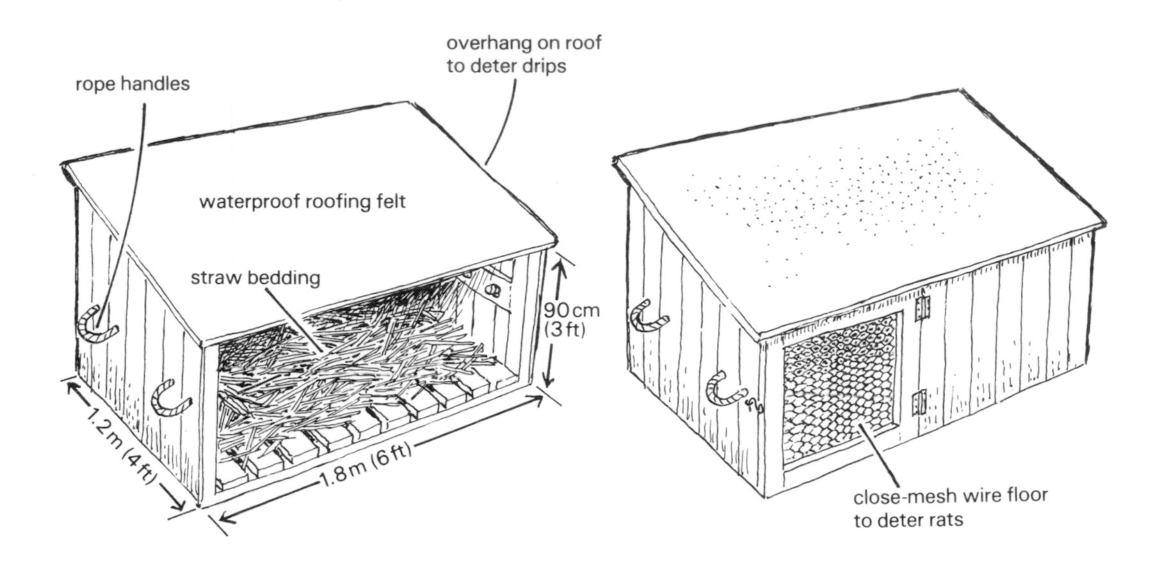

A simple night-time or bad weather shelter for three or four geese

The same shelter adapted as a breeding pen

and, once the geese have finished laying and are starting to sit, a door to confine them.

YOUNG STOCK

In February the geese will begin to lay, and if the breeding sets have been successfully established and mating has occurred, the eggs will probably be fertile. It is always a good practice to take away the first few eggs that a goose lays, in case these are not fertile, and to let her establish her routine of laying in the same place each time. Hopefully this will be in one of the breeding pens set up in the orchard. The door should be left open to allow access in and out and each time the goose lays an egg she will cover it up with straw before leaving. Sometimes, more than one goose will lay in the same nest. This does not matter as long as more than one goose is not allowed to sit and incubate on the same nest. This can lead to problems, particularly that of insufficient incubating, where one goose lays eggs after the other bird has already begun to sit on hers, and in the ensuing mix-up of eggs, some will hatch early and the others may then be left to go cold. There is also an increased risk of eggs being trampled and broken.

There is no guarantee that a goose will use the nesting place provided for her. Sometimes birds ignore pens in favour of a pile of dead leaves in a hedge. If all else fails, it may be necessary to build a temporary shelter around a contrary goose, rather than try to move her.

The incubation period can be anything from thirty to thirty-five days, depending on the type of goose and size of the egg. During this time the goose will turn the eggs several times a day and will emerge in order to feed and drink. It is important that there is sufficient depth and volume of clean water available for her to totally immerse her head and splash her feathers. When she goes back to the nest her damp feathers help to maintain the humidity of the eggs. A few days before hatching occurs you should confine the geese to their breeding pens and provide food and drink inside the shelters, so that the risk from rats, weasels (or even mink in some parts of Britain) is reduced.

It is a matter of individual preference whether the goslings are left with the geese or taken away and placed in a protected environment. It is certainly more convenient to let the mothers do the initial caring, until the goslings are relatively hardy, but again, the risks from predators cannot be overemphasized. One method is to fence off the area immediately around the breeding pens so that the geese and their offspring are restricted to a small area of land. Better still is to take the day-old goslings away from the mothers and keep them indoors, under an infra-red lamp, as would be the case for day-old chicks. With this method there are far fewer losses.

Chick crumbs should be available to them at all times so that they can help themselves whenever they feel like it. Fresh, clean water should be placed in containers that will allow immersion of the beak, but which will not let the goslings climb in. They have a great tendency to paddle into water and can easily become chilled if the feathers are soaked. For this reason, suspended containers are preferable to those which are on the ground.

Some people prefer to incubate goose eggs artificially and there is no reason why this should not be carried out if an incubator is available. The same procedure as outlined for chickens' eggs should be followed, bearing in mind that a slightly higher degree of humidity is required. For this reason, it is important that the instructions for the individual incubator are followed. If you have no instructions for your incubator, spray the eggs with water from a small garden sprayer once a day to maintain the humidity level.

Geese are good mothers, but are not effective at protecting the goslings against rats and other predators.

FEEDING PRACTICE

As already indicated, there is really no substitute for chick crumbs to get the goslings off to a good start. It is a balanced food and, together with free access to clean drinking water, will provide all their nutritional requirements. From the age of three weeks onwards they can be given growers' meal twice a day, with the changeover taking place gradually over three days, so as not to be too abrupt. At this point they can also start to graze on clean grass. If preferred, wheat can be given instead of growers' meal. The supplementary feeding should continue for about another four weeks, with two meals a day being given in the first two weeks (morning and afternoon) and this being reduced to the afternoon meal only for the second two weeks. It must be emphasized again that the grass should be fresh and ungrazed by poultry, because of the high risk of infection by parasitic worms from previous flocks. The main cause of trouble in young geese, which produces a condition known as 'going light', is the gizzard worm.

Once the grass begins to decline at the end of August supplementary feeding will be required again. From the middle or end of September onwards (depending on the grass) the Christmas geese should be given a fattening ration. A convenient and effective one is made up as follows:

1 part barley
1 part wheat
1 part oats or maize (corn)

If any skimmed milk is available from your own dairy animal, the barley meal can be soaked in a little of this to give what is called a good 'finish'. The ration is given twice a day and any spare leafy vegetables, such as lettuces or cabbages that have gone to seed, can also be thrown to the birds. Some people prefer to move the fattening stock from their grazing area to a more confined fattening area for the last ten weeks before slaughtering. This is a matter of individual preference, but it does provide an opportunity for cleaning and liming the grass, as detailed in the section on maintenance of pasture. Keeping the stock in straw yards at this stage was a traditional practice, and may be appropriate for anyone with access to large quantities of straw. Geese are messy birds and their litter will need frequent attention.

Breeding stock should, of course, be kept separate and, as previously mentioned, an orchard is a suitable place for them to overwinter because of the availability of windfalls as a supplementary feed. This will not provide all their nutrients, and wheat should be given, with a certain proportion of oats if available. A good combination is:

3 parts wheat
1 part oats or maize (corn)

Winter vegetables, such as brassica leaves, can also be given, and in the USA, alfalfa hay is a good feed source.

It is not a good idea to let fattening stock graze under fruit trees because they may eat so many windfalls that they are not interested in other feed and therefore will not fatten effectively.

WATER

People often ask whether geese need access to water for swimming. They do, of course, need water for drinking at all times, but they do not need to swim. The important thing is to provide a sufficient depth so that the head can be totally immersed, and water can then be thrown back over the feathers. Unlike ducks, which in the wild, forage for a large proportion of their food in ponds, geese are predominantly grazers and therefore spend most of their time on grass. In winter, it is particularly important to ensure that the water is kept free of ice, and this usually involves breaking the ice several times a day.

Health

Geese are renowned for being healthy and hardy, and provided they are fed adequately, have clean pasture and housing and are looked after while young and vulnerable, there is no reason why problems should arise. When a problem does appear, it is usually that the young geese 'go light'. This is a direct result of grazing pasture infected by gizzard worm, and the importance of land rotation cannot be over-stressed. A vermifuge, such as thibenzole, dissolved in their drinking water will dispose of the worms, but unless the birds are moved to fresh ground, they will quickly become infected again. Adult birds can tolerate a certain level of infestation, although it weakens them, but goslings quickly succumb and will die in a matter of days unless treated. It is important to keep an eye open for any bird beginning to slow down, spending much of its time sitting, or which has a huddled appearance. The local veterinary surgeon is the best one to advise on which vermifuge to use and on what quantities to dose. It is a good policy to worm breeding stock regularly, preferably every three months, to prevent an indigenous build-up in the flock.

Geese can become affected by aspergillosis, colds and pneumonia, bumblefoot and a number of staphylococci infections which have already been detailed in the section on chickens. If feed and drink containers are kept clean, and

Geese fit in well with larger livestock, which eat down the longer grass making the shorter growth more readily available for them.

Picking up and carrying a goose

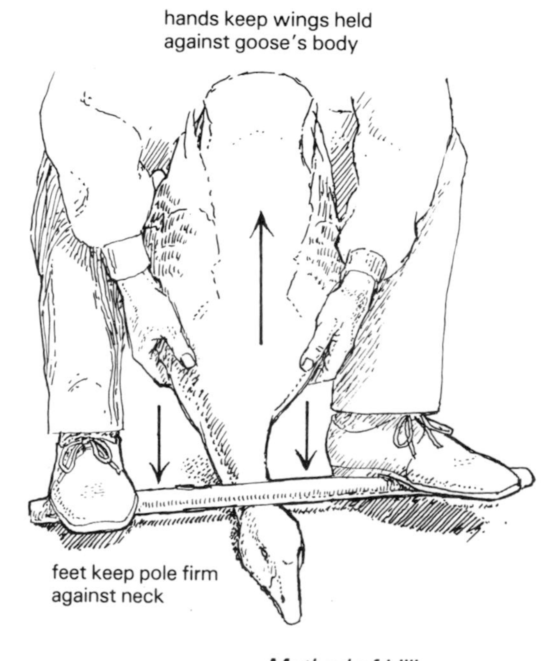

Method of killing a goose

dirty litter is removed from housing, these conditions should rarely appear. A problem which is more common, however, is that of a 'slipped wing': either one wing is held lower than the other, or the last digit is actually sticking out at a right angle to the body. It is the result of muscular weakness, and if a number of cases are encountered, it may be the result of inadequate feeding leading to vitamin and mineral deficiencies. Alternatively, it may be a genetic defect and for this reason, no bird affected in this way should be kept as breeding stock. If treated early enough, it may be possible to improve the condition slightly, by bandaging the wing into place, although the wing must be released for exercising every day.

Killing geese

Geese are more difficult to kill than other poultry and it is best for two people to carry it out, as their strong wings can inflict injuries. When catching and picking up the bird it is essential to keep the wings confined, as shown in the diagram. To kill the bird, one person holds the neck extended along the ground and the other places a pole, such as a broom handle, over the neck. The feet are then placed on the pole, on either side of the head, and the body is jerked upwards. Alternatively, the goose can be stunned on the back of the head and the jugular vein cut at the base of the bill. Some people prefer to chop off the head, using a sharp axe and a wooden block. Whichever method is used, there must be no pain or distress caused to the bird and death must be immediate.

Plucking should follow straight away, using any of the methods detailed for chickens or, if the feathers are required, the procedure described for ducks.

The goose year

The goose year starts in the autumn with the formation of breeding sets, continuing with production of goslings in early spring, followed by the grazing period, fattening and slaughter, and back to over-wintering again. This is indicated in the following sequence, where an orchard is used for the adults.

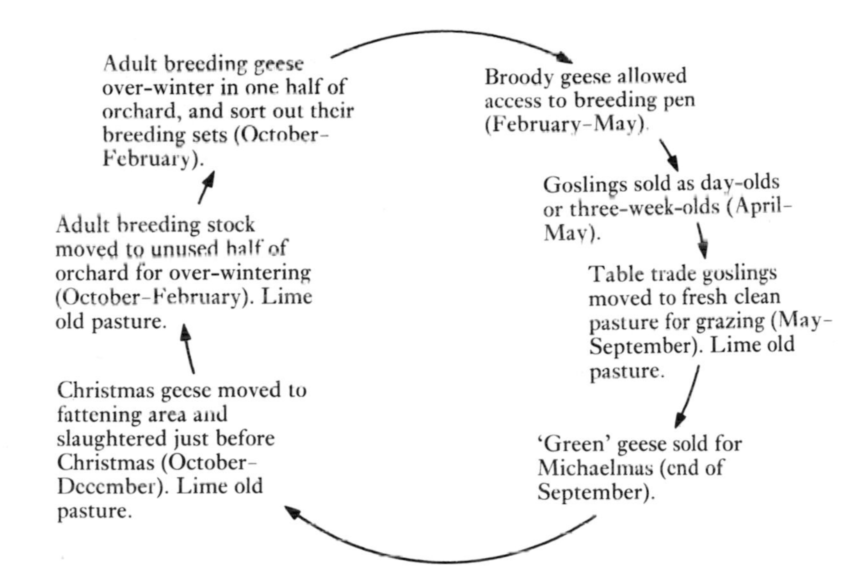

Geese can live to a great age and will continue to breed well for many years. In order to avoid confusion, use some means of identification. The best way of doing this is to place coloured leg-rings on the birds. These are also available as numbered rings so considerable variation is possible. The rings are essential if selective breeding is being carried out.

14 TURKEYS, QUAIL & GUINEA FOWL

Turkeys

Turkeys grow very quickly, and if suitable housing is available and there is sufficient local demand at Christmas or Easter, they are well worth rearing. However, it is not worth trying to breed them, as the newly hatched poults are vulnerable to a variety of infections, including colds, chills and the notorious blackhead condition. The latter is now controlled by a rigid programme of inoculation, but even so, turkeys must be kept quite separate from other poultry because of the risk of infection. Natural incubation is also difficult because the hen bird is a notoriously bad mother and, in fact, most turkey poults are artificially incubated and reared. So, the best approach is to forget about very young turkeys, buy them in as six-week-old poults from a commercial breeder when they are hardy, and raise them to table weight for Christmas.

The turkeys available from breeders are a far cry from the traditional breeds, which were slower growing and smaller. The modern hybrid strains have been developed for quick growth and an efficient feed conversion ratio. They are normally sold as A/H stock, meaning 'as hatched'. In other words, they will be a mixture of males and females. The males (stags) grow more quickly than the females, and also

Stag and hen birds of a modern hybrid strain, which are quicker growing than the traditional breeds, and have an efficient feed conversion ratio.

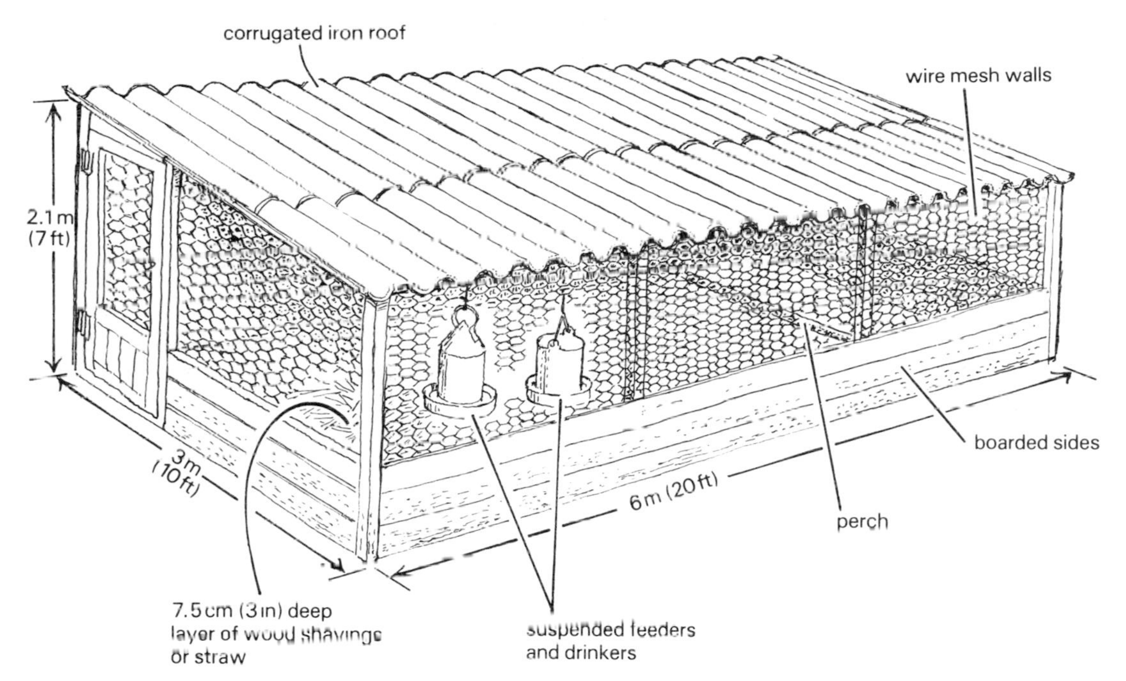

A small home-made turkey house

grow bigger. For this reason, the large commercial producers of oven-ready birds separate the sexes at about six or seven weeks and give them slightly different rations, the hens being given a diet with a reduced protein level to correspond with their slower growth rate. However, this difference is marginal, and the small producer should not bother to keep them separate.

HOUSING AND MANAGEMENT

Traditionally, turkeys were raised outdoors on sandy soil where the efficient drainage was a protection against the blackhead protozoans becoming established. Occasionally one still comes across small numbers being reared in this way, but most small farmers now keep them indoors in a 'pole-barn' system. This system, using a simple barn structure of poles supporting a roof, was pioneered in the United States. The walls are open to the air, although wire netting is used to confine the birds. This method has much to recommend it, and has certainly never been bettered as far as the small producer is concerned. The structure provides shelter, but allows plenty of natural sunlight in through the sides and adequate ventilation. Once turkeys reach the age of six weeks they have become extremely hardy, belying their poor stamina for the first few weeks of life, and respond well to the conditions of a pole barn. In areas where high winds are prevalent, or in particularly exposed situations, the side facing the prevailing wind should have a solid wall.

Wood shavings to a depth of 7.5 cm (3 in) provide an efficient absorbent layer for droppings. Alternatively, straw can be used. Perches are popular with the birds, as long as they are not too high. As turkeys have been bred for increased size, they are not as mobile as their ancestors, who were tree dwellers, and sometimes find difficulty in jumping upwards. Nevertheless, they still have the instinct to perch and are more content if they are given the opportunity of doing so. Suspended feeders and drinkers are best because litter is not scratched into them. They can also be progressively raised as the turkeys grow.

It is important not to panic turkeys. They do have a tendency to flap, and once one of them starts, the rest join in, and before you know where you are, there is pandemonium. The danger here is that they will hurt themselves, either bruising each other or even drawing blood by banging their wings against the walls. The golden rule is, therefore, to approach the turkey house in a quiet, calm way, with no sudden movements. They do get used to the same person feeding them and there is less tendency to hysteria if they see someone they know.

FEEDING

As they are such rapid growers, turkeys do need a relatively high protein level in their food. When they first arrive at six weeks of age, they will probably still be on a starter crumb feed, which is the equivalent of chick crumbs. They can continue this for about a week, so that the change of environment is not aggravated by a coinciding and abrupt change of diet. After this, they can go over to a turkey rearers' ration up until two weeks before killing, when they switch to a finisher ration. All these specialized feeds are available from feed supply merchants. They need to be fed three times a day, ideally first thing in the morning, at lunch-time and in the afternoons, with enough being given to keep them eating for twenty minutes. If they finish before this, it is an indication that they are not receiving enough – and vice versa.

Many people prefer to feed a less intensive diet, not only because it is cheaper, but also because they are concerned about the antibiotics which are included in the rearers' ration. This, as previously mentioned, is to combat the risk of blackhead infection. One alternative is to give one feed of rearers' ration in the morning, then a half-and-half mixture of wheat and oats at lunch-time and in the afternoon. In addition to this, feed any available kitchen scraps and vegetable peelings. These, however, must be chopped up small, otherwise there is a danger that the digestive system will become blocked. If grain is given, it is also important to make oystershell grit available so that the grain can be digested.

Traditionally, growing turkeys were given the following mixture:

1 part bran
2 parts maize meal (corn)
1 part fish meal
2 parts wheat

It is important that if such a diet is given, the fish meal allowance is not exceeded, otherwise the meat may acquire a fishy taste. It is advisable to stop giving fish meal two weeks before killing anyway.

KILLING

The turkeys will be killed just before Christmas, but the precise date will depend upon the individual. Bear in mind that enough time must be left to kill, pluck and draw the birds, and to leave them to hang in a cool room for a few days. There is no point in killing them so early that buyers have to freeze their birds to stop them going off. The whole point of the exercise is to provide people with freshly killed turkeys; they can get frozen ones in the supermarket.

Ideally, the turkeys will have been ordered beforehand, and the usual practice is to record the requests as follows: Mrs Brown: 5.4–6.8 kg (12–15 lb). It is important to abide by these requests. If Mrs Brown has ordered a turkey within a certain weight range, she has good reason, and she will probably not be happy with a 4.5 kg (10 lb) bird, which will not feed her family, or a 9 kg (20 lb) one, which will be too big for her oven. This may seem an obvious statement, but it is surprising how often a situation like this does happen, and the customer may go elsewhere next Christmas.

The birds should not be fed for at least twelve hours before killing, although water should be freely available. They are killed in the same way as previously described for chickens, and plucked immediately. As there may be a large number to do, help may be needed with the plucking. Friends and family are often ready to help in return for a turkey, or if the enterprise warrants it, it may be necessary to pay casual labour by the hour. There are usually experienced pluckers in country areas who would welcome the extra income before Christmas.

Turkeys are gutted in the same way as chickens, although the leg tendons are much stronger and will need special attention to get them out. Some buyers, of course, will not require their birds to be gutted, and the usual practice is to kill, pluck and hang the bird with a label tied to it showing the buyer's name, the weight of the turkey, the price per pound and the price of the bird itself. Turkeys sold gutted and dressed will obviously command a higher price than those which have merely been plucked. The regulations which apply to the sale of turkeys are the same as those applying to the sale of chickens.

Once the turkeys have gone, the house should be thoroughly cleaned and fumigated before being put to use again. This could be to raise more turkeys for Easter, or possibly ducklings, which also sell well at this time. Alternatively, broiler strains of chicken could be raised there, but it goes without saying that the cleaning and fumigation processes must be thoroughly carried out between batches of poultry.

Quail

The keeping of quail is a minority pursuit, but if you are already interested in poultry, it is worth examining the possibilities of quail to supply the small but growing delicatessen market for quail meat and eggs. As with other poultry, there are different strains for egg laying and meat production, and it is important to acquire the right strain for the right purpose. A commercial strain of the Japanese quail (*Coturnix japonica*) is the best for egg production; a good strain can produce up to 300 eggs a year under the right conditions. If kept for the table, the Bob White variety (*Colinus virginianus*) is appropriate because it is heavier and puts on weight more rapidly than other breeds. Commercial strains of both types are normally advertised in the poultry press.

As far as regulations are concerned, the same rules apply to quail as to chickens, and as long as the small producer is supplying local consumers, either direct or via local retailers, and is mindful of the health and humane regulations, he will have no problems.

HOUSING

The biggest enemy of the quail is the rat, and the first priority for housing is that it should be totally rat-proof. Normally, adult poultry can cope with and dismiss a rat, but bear in mind that the adult quail is only 18 cm (7 in) long. Commercially, quail are kept in runs with wire floors, divided into separate compartments, with a pair of birds in each compartment. The average size of each compartment is 45 cm × 1.1 m (1 ft 6 in × 3 ft 9 in) with a nesting area on one side, the other side being equipped with a feeder and drinker.

Rabbit hutches have been used to good effect but the floor should, in this case, be kept covered with a layer of clean litter such as wood shavings.

The housing should be kept in an outhouse where electric lighting is available, otherwise the number of eggs will be reduced in winter. Some people do keep quail in a small poultry house with attached run, but it must be emphasized that they are in danger from rats in this situation. I once kept some of my older breeding stock this way, but even though I went to the trouble of placing wire mesh under the turf in the run, the rats still got in and took the whole stock. They had obviously found a weak point in the wire mesh and gnawed through it. Let no-one suppose that their land is free of rats. Wherever livestock is kept, there will be rats, and although a great deal can be done to keep down the numbers, total eradication is impossible. (See page 32 for information on dealing with rats.)

A convenient way of housing stock is to use a series of adapted rabbit hutches in an outbuilding equipped with electricity. In order to ensure that the light reaches inside the hutches, the roof of the main part is removed and replaced with wire mesh, while the roof of the sleeping compartment is retained. Part of the floor can also be treated in this way so that cleaning out the hutches is simplified; the droppings are merely brushed onto the mesh and then fall through. Feeders which can be filled from the outside are the most convenient, and the hopper type normally used to feed rabbits, are suitable. Similarly, drinkers which can be

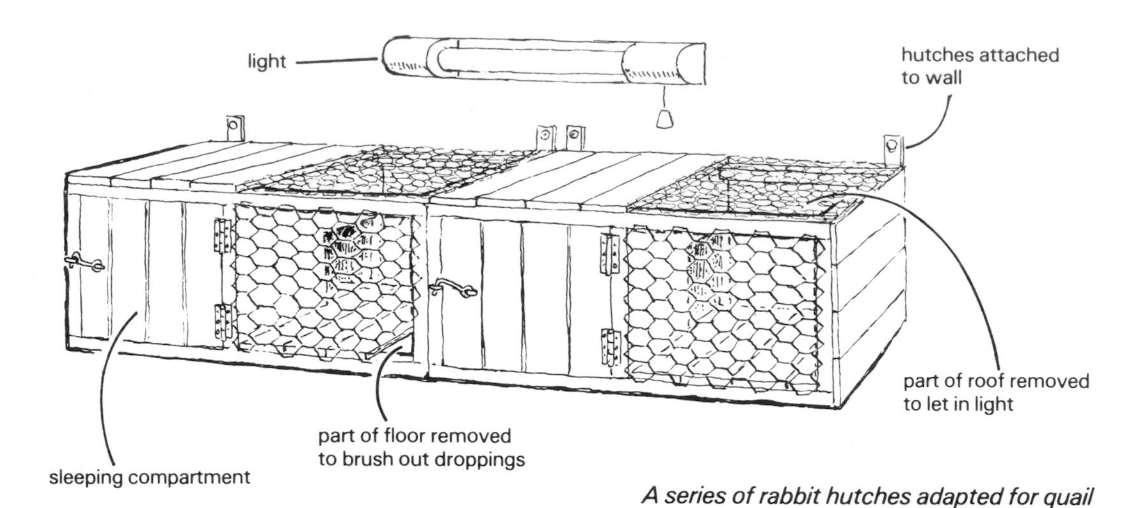

A series of rabbit hutches adapted for quail

refilled without the hutch having to be opened are best, and again, a gravity-fed rabbit drinker is suitable.

FEEDING

It has been estimated that thirty quail can be kept for the same price as one chicken. They do best on a high protein ration and proprietary chick crumbs are suitable for all ages. This is best supplied on an *ad lib.* basis where the birds can feed at will. An alternative is to mix three parts chick crumbs with one part budgerigar seed (available from pet shops). If grain is included in the diet, grit will also be needed to aid in its digestion. Chopped lettuce leaves and those of leguminous plants, such as peas, are also popular. Water should, of course, be available at all times.

QUAILS' EGGS

Young quail will come into lay at about the age of ten to twelve weeks. The eggs are small, weighing less than 14 g ($\frac{1}{2}$ oz), and are distinctly patterned in brown, black or blue on a white background. Some strains do, however, lay all-white eggs, although these are fairly uncommon. Each bird tends to lay its own distinctly patterned egg so that it is quite easy to spot any non-productive hen birds.

The egg-laying strains will produce approximately 300 eggs a year, with most of them being laid in the first eight months of production. After this, the numbers decline and the birds should be replaced. The numbers will also decline in winter unless electricity is provided to give a minimum of fifteen hours of light a day.

The eggs can either be sold fresh, by the dozen, or as the delicacy 'quails' eggs in aspic'. Egg containers provide a difficulty because the eggs are too small for the normal ones. No one has yet come up with a suitable quail egg container, and it is a matter of selling them in paper bags and handling them with scrupulous care.

Aspic jelly to cover hard-boiled quail eggs is easy to make. You can either buy the ready made product, to which you need to add boiling water, or you can make your own. A simple recipe is to take strained, clear stock and add gelatine to it in the proportions of 28 g (1 oz) gelatine to every 1.3 l ($2\frac{3}{4}$ pt) of stock. Stir well, then heat to boiling point. When it begins to cool, pour the jelly over the hard-boiled eggs (from which the shells have been removed) and leave to cool and set. Any appropriate containers, for example waxed paper trifle cases, can be used. Take care not to over-boil the eggs, as this will make the shells difficult to remove. They should not be boiled for longer than thirty seconds.

For general cooking purposes, three quail eggs are equivalent to one chicken egg.

QUAILS FOR THE TABLE

The commercial table strains of quail are normally ready for killing at six weeks, when they will have increased in weight from about 14 g (½ oz) at hatching to 225 g (½ lb) at killing. During this time, approximately 4.5–4.9 kg (10–11 lb) of feed will have been consumed.

Killing is by neck dislocation, as for chickens, and plucking and gutting are also similar. Special care needs to be taken to ensure that the skin is not damaged by over-vigorous plucking. The meat is light brown in colour and is regarded as a delicacy. An easy way of cooking quail is by roasting, wrapped in bacon slices, and serving with any of the sauces and garnishes normally used with chicken or game.

BREEDING REPLACEMENT STOCK

Quail are not reliable at hatching their own eggs and it is better to incubate them artificially. Where a broody bantam hen is available, she will be able to take up to twenty eggs, depending on her size. If a quail does become broody, she can sit on eight eggs. The incubation period is seventeen days, but the chicks may hatch out a day earlier or later. In an artificial incubator, the manufacturer's instructions should be followed to ensure that the correct degree of humidity is achieved. Most modern incubators will include specific details on quails' eggs. If you are using an old incubator, where instructions are no longer available, ensure that the temperature is at 100°F (37.7°C) and that there is a container of water, which is kept topped up to ensure that the air does not dry out. The eggs should be turned five times a day.

Once the young have hatched they will need to be brooded in a protected environment, and an arrangement such as that detailed for chicks is suitable (see page 135). There will need to be certain adaptations, however, to cater for the much smaller quail. The newly hatched young are really tiny, not much bigger than bumble bees, and great care has to be taken to ensure that there are no spaces for them to squeeze out of their protected environment. The other important point is to place pebbles in their drinker, to decrease the depth of water, so that they do not drown.

They can be given proprietary chick crumbs from the start, and do well on this high protein diet.

Guinea fowl

There is a small, gourmet market for guinea fowl with its delicate and distinctive game flavour. Even the Greeks and Romans favoured them for the table, having introduced them from the Guinea coast of Africa. The question is whether you can put up with the noise they make before they reach the table. Guinea fowl are members of the pheasant family and are extremely nervous, flighty and loud. The females in particular will shriek if they see someone coming, while the rest keep up an almost continuous warbling. They undoubtedly make good watchdogs, and some people keep one or two for this purpose, but the proximity of neighbours is a factor to take into consideration. Guinea fowl which are kept as watchdogs will sleep in trees or on top of high buildings, and will spend a great deal of time searching for insects, which they relish. In fact, they are extremely efficient at clearing land of insect pests and are often put to work in this way after the harvesting of crops.

PRODUCING TABLE BIRDS

If guinea fowl are kept for the table, they are best kept confined, either in an adapted out-

Guinea fowl make good watchdogs and will clear the ground of insect pests, but their noise may bring complaints from neighbours.

building, such as that used for turkey rearing, or using a house and attached straw yard system like the one detailed for the rearing of table chickens (see page 136). Where a run is used, the birds will need to have their wings clipped, otherwise they will fly straight over the fence.

It is important to rear commercial hybrid strains rather than the ornamental varieties because the former put on weight more rapidly and require less feed consumed in order to do it. When housed indoors and fed on growers' rations, guinea fowl will be ready for killing around nine weeks of age, when they will have achieved a weight of about 900 g (2 lbs). If they are allowed outside, they may take up to fourteen weeks to reach killing weight.

FEEDING

Proprietary chick crumbs on an *ad lib.* basis, together with fresh, clean water, will provide for all the nutritional needs of the young guinea fowl, which are referred to as 'keets'. After the first three weeks they can be switched to a diet like that suggested for the rearing of table chickens. The thing about guinea fowl, however, is that they have small crops and therefore feed more frequently, taking less volume at a time than other fowl. This can lead to problems of effective feeding unless the feed is available on an *ad lib.* basis, and for this reason, many people find it more convenient to feed proprietary growers' rations, rather than making up the rations themselves. Another important factor is that some broiler rations contain coccidiostats which are toxic to guinea fowl. It is worth consulting your local feed supplier on this point. Some people have successfully reared guinea fowl on proprietary pheasant rations, which are normally available from game suppliers.

15 RABBITS

There is no doubt that rabbits are among the easiest of livestock to keep. They are quiet and do not require expensive housing, unless they are being kept on a massive scale. A breeding trio of one buck and two does will provide a family with sufficient young table rabbits through the year. If a commercial unit is envisaged, the scale would need to be bigger, with at least twenty-five does and two bucks. This size of operation would ensure a sufficient and regular turnover of table rabbits for sale.

Commercial rabbit packers are those who are responsible for the collection and distribution of table rabbits in Britain. Where the number of rabbits is high, they will collect the live animals direct from the farm. With smaller numbers, however, this facility does not exist, and as a result, smaller rabbit producers have got together to form local groups which act as central collection points. The beginner is able to obtain much useful advice and information about local conditions from such a group, and a first step would be to make contact. The local library should be able to help with names and addresses.

The Commercial Rabbit Association is the organization which looks after the interests of the commercial rabbit producer in Britain. It offers much useful advice, including a list of accredited members who will supply good quality breeding stock. The body to contact in the United States is the American Rabbit Breeders' Association.

Suitable breeds

There are really only two breeds to consider for a commercial enterprise, the New Zealand White and the Californian. Both developed in the USA, they have whiter meat and a higher meat to bone ratio than the more traditional breeds. The New Zealand White is slightly faster growing, but the Californian has a thicker pelt.

There are many commercial breeders

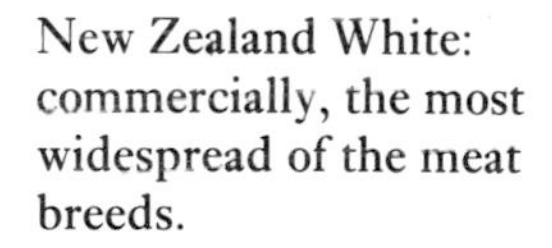

New Zealand White: commercially, the most widespread of the meat breeds.

Californian: one of the best meat breeds, and has a good pelt.

supplying good quality breeding stock, as well as a range of housing and equipment. It is much better to obtain stock from one of these, than from an amateur breeder, for, although more expensive, the animals will be guaranteed free of disease and defects. The Commercial Rabbit Association has a list of accredited suppliers. Ask whether the rabbits have been inoculated against myxomatosis. Often, this is available as an optional extra. Alternatively, the veterinary surgeon will inoculate them, or you can do it yourself once you have been shown how to by an experienced person.

Housing

You will have to decide on the system of housing before acquiring any rabbits, and the choice is between cages and hutches. Cages need to be housed in a building for weather protection, but it does mean that electric lighting in the building is possible with this system. Without this, the rabbits cannot be successfully mated throughout the year.

Cages are also easier to keep clean because the droppings fall straight through onto straw-covered ground, which absorbs them. This litter should be replaced periodically to avoid a build-up of ammonia fumes from the urine, which could cause eye irritation to the rabbits. In some units straw is not used, and the concrete hosed down every day.

Some people prefer to use tiered hutches, feeling that cages are too reminiscent of the battery hen system. The cages, however, are much more spacious than those used for poultry, and the rabbit itself is much better adapted to confinement than is a hen. More cleaning out will be needed with hutches and electric lighting poses problems. This can all be overcome, of course, and some rabbit-keepers have compromised between hutches and cages by using single-tiered hutches with a set-in droppings panel. Each cage will need to be supplied with a feeder and drinker, and an automatic watering system is by far the most convenient. These are supplied by most commercial rabbit stockists.

Feeding

Commercially, proprietary pellets are fed, which, together with water, provides all the nutrients needed. They are expensive, however, and the fact that they contained antibiotics to combat coccidiosis and to promote more effective growth, is worrying to many people.

Breeding stock is fed an average of 115 g (4 oz) of pellets a day, while pregnant does are given 225 g (8 oz) a day from the twenty-fourth day of pregnancy, and throughout the lactation period.

Young rabbits, after weaning, are fed on an *ad lib.* basis, and are given as much as they will take until slaughtering time.

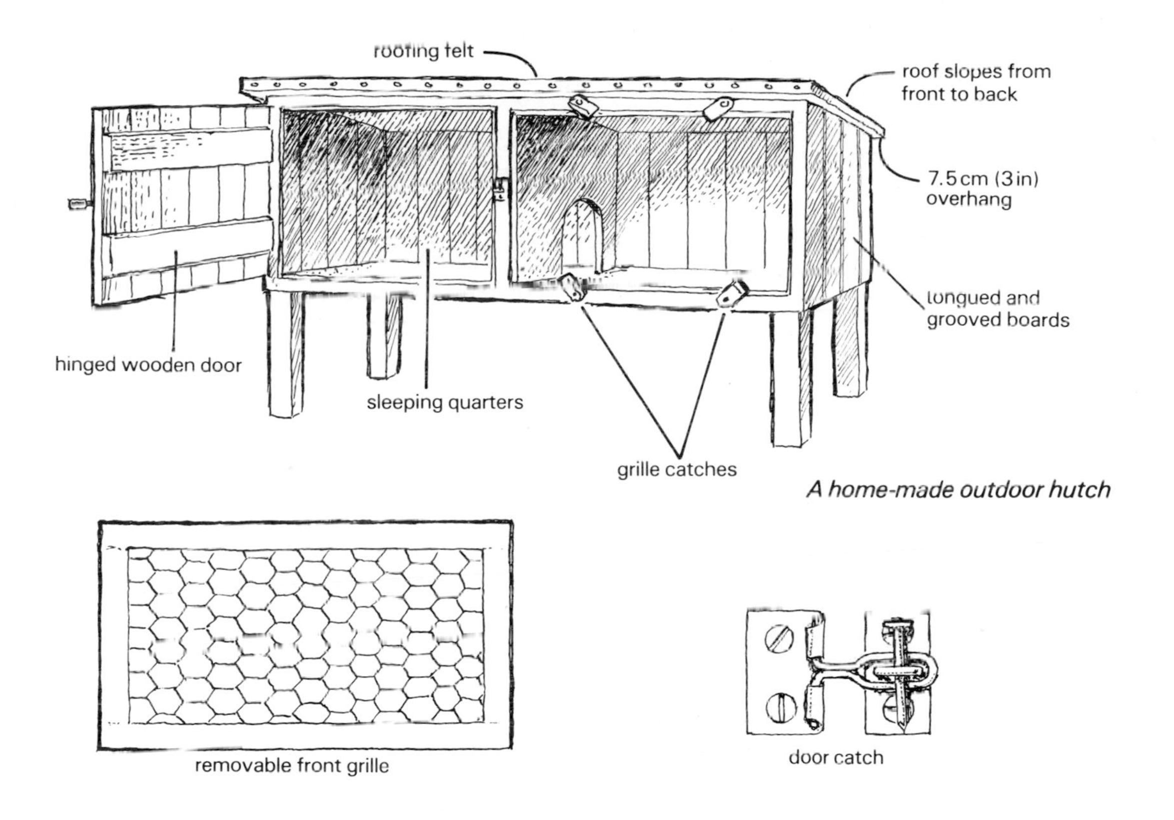

A home-made outdoor hutch

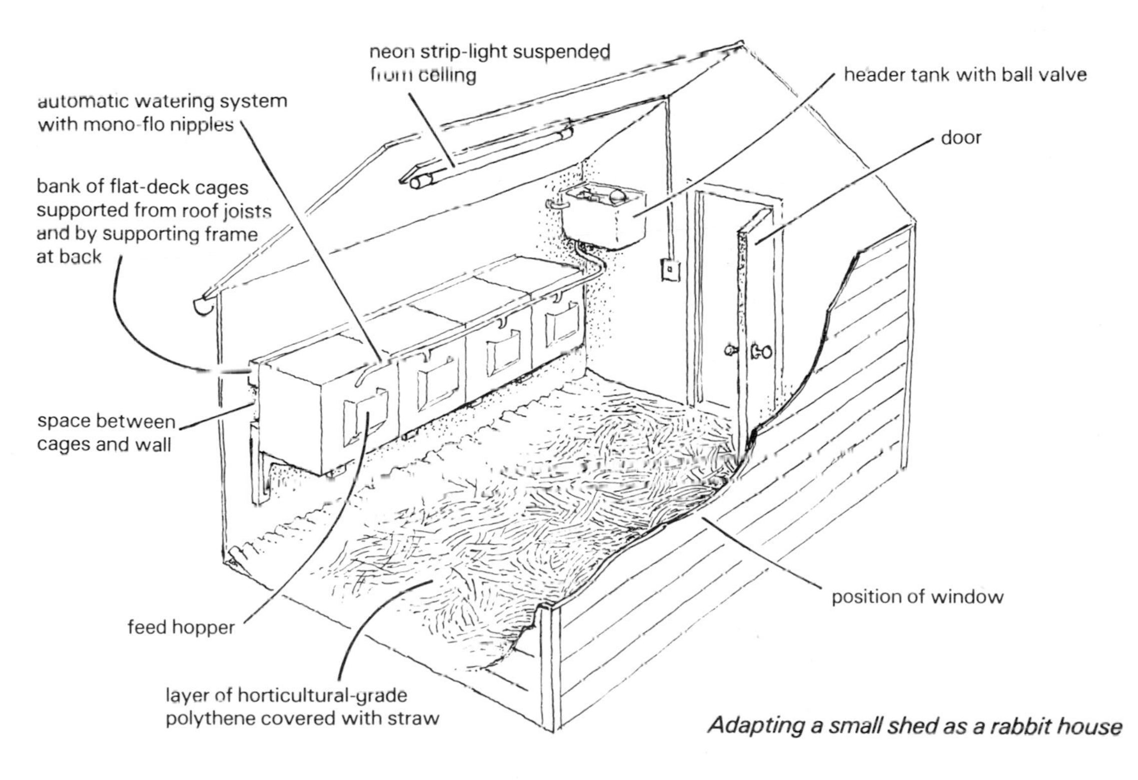

Adapting a small shed as a rabbit house

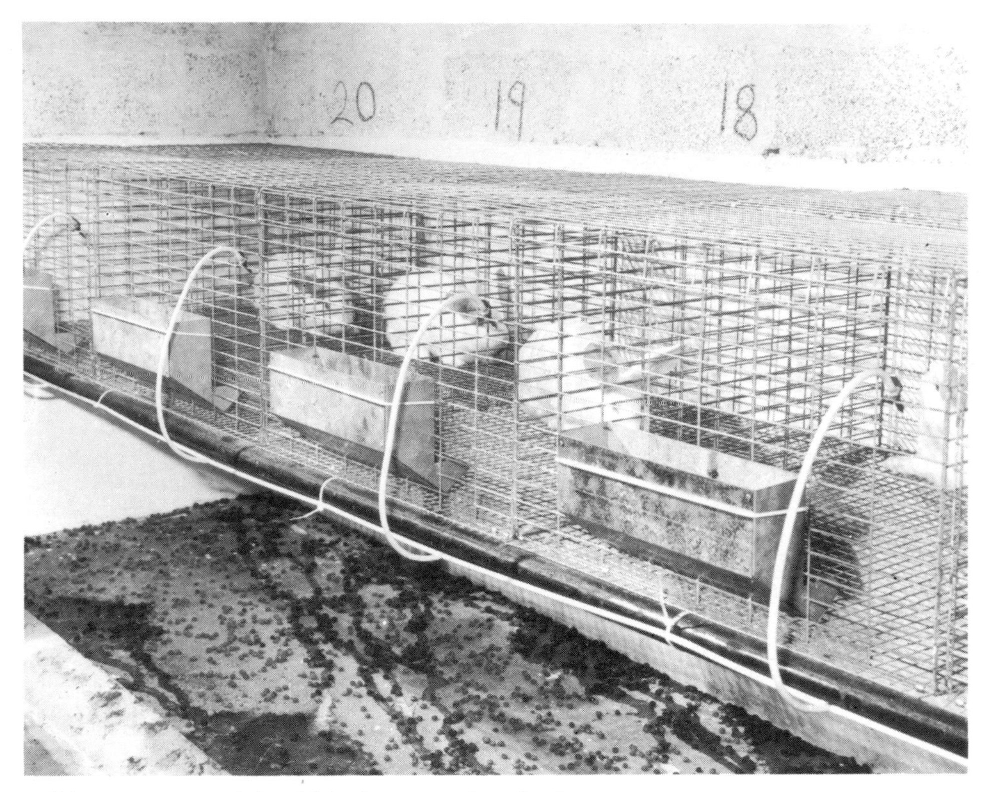

Many commercial rabbit farmers also feed hay, in small hay racks inside the cages, on the basis that the fibrous material is more akin to the rabbit's natural diet, and that digestive problems are therefore avoided. This is provided once a day so that the rabbits can help themselves when they feel like it. It is, of course, possible to feed less intensively, although this is more time-consuming. Many people are anxious to use a proportion of home-grown feedstuffs to reduce costs, and utilize hay, oats, bran, crisped potato peelings and garden greens and roots. Take care to ensure that all the nutrients needed are being supplied, and that digestive problems are avoided by not giving too much of any one particular green food. Many people feed a proportion of pellets and make up the rest with hay and the home-produced feeds, in an effort to reduce costs, and this works quite well, but the time spent collecting and feeding feedstuffs other than pellets should not be underestimated. This is a decision that you will have to make when you have considered all the alternative combinations, and the time element.

Breeding and general management

When breeding is required, the doe is put in the buck's cage, never the other way around or she will attack him. Usually, mating will take place within a matter of minutes, but sometimes it happens that the doe runs about and will not stand for him. In this situation the best solution is to conduct an 'assisted mating'. This involves holding the doe in a stationary position, with one hand underneath her abdomen, so that her hocks are slightly raised, while the other hand holds the scruff of her neck so that she cannot escape. In this way, mating will occur quickly. Both buck and doe should be at least twelve weeks old before they are used for breeding. They should obviously be in perfect health, and free from physical deformities.

Depending upon the number of does that you have, you will need to work out approximately how many litters per year you require, so that you have some idea of the total number of table rabbits you can expect in a year. In a commercial enterprise, each doe produces an average of six litters a year with a minimum of nine per litter aimed for. Any doe which produces only small litters should be replaced by another breeding doe.

The keeping of records is vital, and each doe should have her own record card attached to her cage or hutch. This will give details of her breed and number, the date and number of the buck she mated with, followed by the date of kindling or giving birth, with the number of live and dead births. Some people identify breeding rabbits by placing a metal tag in one of their ears, but this can lead to problems if the tag should catch on the wire of the cage or hutch. It is better to tattoo the number inside the ear.

The doe's pregnancy will last thirty-one days, and during this time she will require a maintenance allowance of 115 g (4 oz) of feed pellets a day. On the twenty-fourth day, however, her feed allowance should be doubled to 225 g (8 oz) to take into account the forthcoming lactation.

If fed on more traditional feedstuffs, she should not have too many fattening foods, such as bread and potatoes, which may lead to overweight and subsequent kindling problems. She should have clean, fresh hay to nibble and some people feed raspberry leaves at this time in the belief that they have a beneficial effect on the uterine muscles during labour. On the twenty-eighth day, if she is in a cage, she should be given a nest box with a little clean fresh hay in it. If in a hutch, she should receive extra hay in her sleeping compartment. She will make a nest in this, and pull out some of her own fur to line it. As soon as you see her doing this, you will know that the birth is imminent. She is best left to her own devices during the actual birth, until the young have all arrived, been cleaned by her, and covered up in the nest. Make a careful examination to check that she is all right, and to count and record the number of young. Any stillborn or physically deformed ones should be removed and disposed of.

Rearing for the table

For the first three weeks, the young will feed on the mother's milk, and although the natural lactation period is seven weeks, the normal commercial practice is to remove the young from the mother at three weeks old. They are then fed on feed pellets and hay. The pellets are given on an *ad lib.* basis until, if the rabbits have come from good initial commercial stock, they will weigh approximately 2 kg (4½ lb) at ten weeks. At this time, they will be ready for slaughter. This is easily done, but you should not try it yourself without having received expert tuition first.

There are two ways of killing, the first using the so-called 'rabbit punch' where a hard blow is delivered to the back of the head, behind the ears. A piece of lead piping or even the edge of the hand will do; the important point is that death should be instantaneous.

The other way of slaughtering is to hold the rabbit upside down, and wring its neck by twisting and pulling simultaneously on the throat. There will be a certain amount of reflex nervous and muscular movement, but this is a normal phenomenon following death. Neck dislocation is preferable from the commercial point of view because there is less discoloration of the skin than with the other method.

16 BEES

A small, but important, proportion of the honey that is on sale comes from the small bee-keeper. There is no problem about selling surplus honey. It can be done through a local bee-keeping society, through local shops or direct to the consumer at the gate. What is more difficult is to make the enterprise pay. The cost of beehives and equipment is now very high and it would take many seasons' honey harvests to recoup the initial outlay. Second-hand equipment is available, particularly through bee-keeping societies, but the novice needs to be careful about purchasing it. The hives may be infected with disease and will need thorough cleaning with a blow-torch to make them safe. In this respect, it is a good idea to contact the bee officer for your particular area, via the local ADAS office. He will provide help and information, and will advise on disease prevention. It is obviously in everyone's interests to avoid an outbreak of disease.

Some people are reluctant to keep bees because they are afraid of them. However, this is usually due to lack of experience. A colony of bees can do little harm if it is handled properly. Anyone who is allergic to bee stings would obviously be unwise to keep bees. No one should go in for bee-keeping without having attended a practical course on the subject, or having made contact with an experienced bee-keeper who can offer advice. Most local authorities run bee-keeping courses as part of their night school programmes, and agricultural colleges, too, have excellent courses. Many equipment manufacturers also run courses, and will offer a great deal of advice to their customers.

Equipment

The hives obviously come before the bees, and it is important to make the right choice for there are now at least six different types. The old

WBC hive

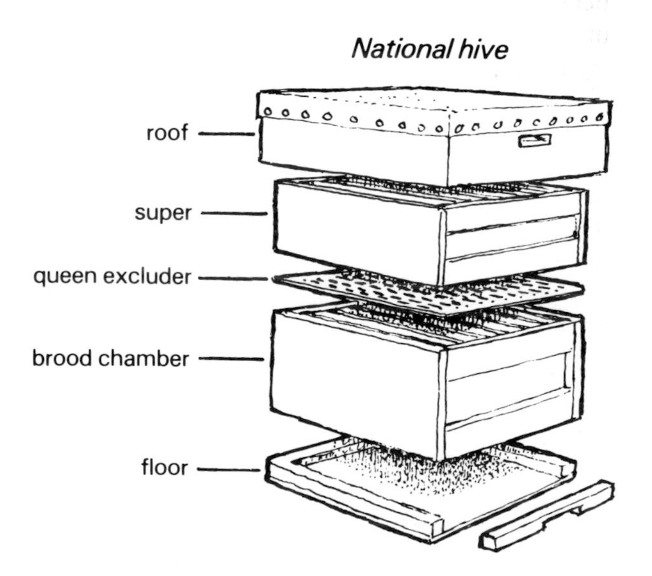

National hive

WBC type looks attractive and it is the one that most people associate with country gardens but it has been superseded by the more up-to-date National hive, although some people claim that it offers better weather protection. The latter is now the most common in Britain, but on a world-wide scale, the Langstroth is the commonest. When buying bees, they are likely to be on British standard frames, which will fit the National and the WBC. The Langstroth and the other types, Dadant, Commercial and Smith, are efficient box-type hives, but are not as easy to move around as the National. Anyone considering the choice of hive should visit one of the many excellent suppliers and view them at first hand.

Other necessary equipment includes a bee veil and hat to provide protection, as well as a pair of gloves. Experienced bee-keepers are often to be found handling their bees without wearing protective clothes, but this is not to be encouraged in the beginner. A smoker is a vital tool, because it does mean that handling the bees is made easier, for the bees become more docile. A hive tool is essential for loosening frames and making them easier to remove from the hive. The propolis produced by the bees tends to make everything stick together and a good metal hive tool is efficient in prising frames apart, without causing damage. Trousers should ideally be tucked into boots or socks because a favourite trick of bees is to climb up inside a trouser leg.

A feeder is necessary for each hive. This is basically a metal container with a tube running up the middle, so that bees can crawl up from the hive. The feeder is placed over the feed hole and then filled with sugar solution. A tube with a glass top is inverted over the central tube so that the influx of bees is controlled. It means that they come up the central tube in small numbers, feed, then return, without being in danger of drowning in the syrup.

An extractor will be needed for extracting the honey from the combs, and this is usually the most expensive item. If you have only one or two hives, it may not be worth buying one, as local bee-keeping societies frequently hire one out to their members. The extractor whirls the honey combs around so fast that the centrifugal force drives the honey out. This then needs to be strained to remove any particles of beeswax. Suppliers of bee-keeping equipment stock purpose-made strainers or a large-sized kitchen strainer can be used. After straining, the honey needs to stay in a settling tank for a day or two so that any bubbles rise to the surface. A plastic tub with a lid such as that used for making beer or wine is ideal for this. Finally, you will need storage jars. Screw-top jars are best, and can often be purchased through bee-keeping societies.

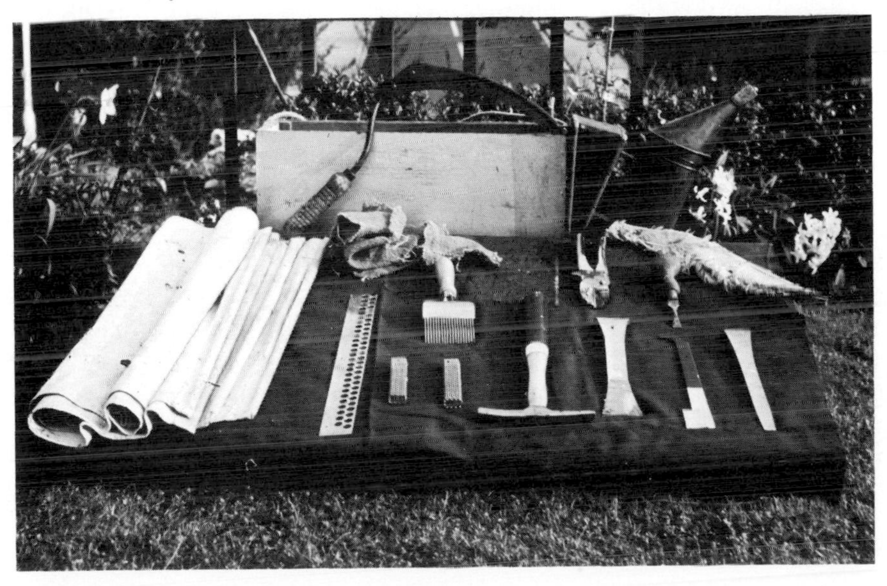

Portable toolbox: *back row* grass-cutting hook, smoker; *centre* smoker fuel, uncapping fork, secateurs, screwdriver; goose-wing bee brush; *front* cover cloths, mouse guard, queen cages, hammer, hive tools.

The hive in more detail

A modern hive is essentially a box with internal ledges on which vertical frames are suspended. It is on these frames that wax combs for egg laying are laid down. For this reason, the box is called a brood chamber, the frames are brood frames and the waxcombs are brood combs.

The floorboard under the brood chamber has one side open to provide an entrance, and this is normally equipped with an entrance block to vary the size of the entrance; in winter it needs to be narrower for weather protection as well as to exclude larger, robber bees or other predators. In front is an alighting board for returning bees. Above the brood box is a queen excluder. This is a zinc or plastic perforated sheet, which will allow worker bees through, but not the queen or the drones. In this way, no eggs are laid in the chamber above the brood box.

This upper chamber is called a super and is designed to hold the honey frames. There may be one or more, and they are similar in construction to the brood chamber, but normally shallower.

Above the supers is a crown board or inner cover, which has a feed hole, over which a feeder can be placed. Covering the hive is a weatherproof roof, frequently fitted with ventilator grilles at the side. There is also a bee escape, which allows bees to go out but not to get back the same way. The only entrance is at the front via the alighting board.

Siting the hives

The ideal place to site hives is in a sheltered, sunny position where prevailing winds do not roar around them in winter. They also need to be away from the immediate area where people are likely to be working.

When bees emerge from a hive, they follow a definite flight path and anyone in the way is in danger of being stung. The way around this is to erect a fence or plant a hedge about 1.8–2.4 m (6–8 ft) away from the front of the hives so that the emerging bees are forced to take a steeply inclined flight path. By the time they are on the other side of the fence, they will be safely above the heads of any unwary gardeners or members of the public. In urban gardens, many people site their hives in a screened-off area at the furthest point from the house. In rural areas hives are placed in an orchard if possible.

Acquiring the bees

Most large suppliers of bee-keeping equipment also supply bees. In addition, there are specialist breeders who may be contacted through an organization such as the British Bee Breeders Association. Local bee-keepers may also be prepared to supply bees, and it is here that a local bee society can be helpful. Some experienced bee-keepers are listed at local police stations as being available to take a swarm of bees in the event of an emergency. In this situation, it is accepted that a bee-keeper who is called out to remove a swarm claims the bees for himself.

Purchased bees will normally arrive in a travelling box and they will need to be transferred from that into the brood chamber of an empty hive. Light the smoker and make sure that it is working well before putting on a bee veil. Trying to light a smoker while wearing a veil can be hazardous, and at least one fatal accident has been recorded as a result of this.

Unscrew the screws that secure the lid of the travelling box, puff a little smoke in, wait a few minutes then gently lift off the lid. Now lift out the outside comb by holding the projecting ends until it is clear of the box. Carefully turn it to examine both sides and check that there is a brood in the middle. Some cells will be capped while others will be unsealed with white grubs in them, and some will have small eggs. Place the frame in the brood chamber of the hive, in the same position that it occupied in the travelling box. Repeat the procedure until all the frames have been transferred. If there are any bees left in the box, turn it upside down and shake them out into the hive. Now place the cover over the brood chamber and replace the roof.

If the weather is fine and there is enough honey already in the combs, supplementary feeding should not be necessary. If it is needed, the feeder should be placed over the feeding

Transferring live bees from a travelling box into a National hive.

hole. The syrup is made by dissolving 900 g (2 lb) of refined white sugar in 0.5 litre (1 pt) of boiling water, then cooling before placing in the feeder.

Keep an eye on the new colony, and as the number of bees increases, provide new frames with foundation wax on them, which can also be colonized. Do this until the brood chamber is full of brood frames. At the beginning of April, put a queen excluder on top of the brood chamber and add a super with foundation frames to encourage the bees to manufacture honey combs. As these become filled, a second super may be needed.

Taking a swarm of bees

Occasionally it may be necessary to take a swarm of bees, either for someone else, or because your own hives have become over-populated.

The easiest place from which to take a swarm is from a low and accessible bough on a tree, but swarms have been known to cluster on the most difficult places, such as roof tops or the underside of lorries. Usually, however, they do tend to settle on trees. You will need a ladder and a pair of secateurs or a pruning saw if it is necessary to cut the bough, and a stout cardboard box for catching the swarm. It is much easier for two people to take the bees: one to remove the swarm and the other to hold the box to receive them. Sometimes a good sharp bang on the branch will be enough to dislodge them but usually it is simpler to cut through the wood. The helper should be ready to take the weight of the swarm when it is severed and gently lowered into the box. People doing this for the first time are often amazed at how heavy a swarm can be

Take the covered box to a nearby hive and gently place it on the ground. Spread a white sheet on the ground in front of the hive so that one side covers the alighting board. In the brood chamber place some brood combs with foundation wax and if possible, include some honeycombs as well. One old custom was to rub the inside of the brood chamber with some

Hiving a small swarm from the skep in which it was collected into a nucleus box, using a cover cloth and mist spray.

broad bean flowers as well. It was claimed that the smell was an added inducement for the bees to colonize it. Place a feeder with sugar solution over the feeding hole in the inner cover and replace the roof. Now gently turn the cardboard box over to tip out the swarm onto the sheet. If you are afraid of disturbing the swarm too much, just turn the box onto its side and leave the bees to crawl out, up the sheet and into the empty brood chamber.

Spring management

It is a mistake to open and examine a hive too early in the season, exposing the over-wintering bees to a sudden drop in temperature. April is normally early enough, and a mild, warm day should be chosen. In these conditions the bees are more likely to be placid, particularly if they have had the advantage of the early blossom.

If all the combs are full of bees and obviously active there is no need to remove any of the frames unless you find it difficult to see from above whether brood cells have been made. Often it is possible to see this by puffing in a little smoke and then looking down when the bees have cleared to one side. If the brood chamber is full, it is best to provide a second brood box or a shallower super to act as a brood chamber. As bees tend to work downwards in the spring, many people prefer to lift up the original brood chamber and put the second brood box underneath. The drawback is that one has to disturb the bees in order to do this.

Ensure that the near brood box has frames with foundation wax. Ideally, combs that have already been built should be used because this lessens the work load of the bees at a time of year when they have a lot to do, but plain foundation can be used if this is not available.

Once the second brood chamber has been placed in position, a queen excluder should be put on top, and a super with frames for the laying down of honey comb. If, on the initial spring examination, the brood chamber has plenty of room for expansion there is no need for a second brood box to be added, and a queen excluder and super can be added straight away. Where the number of bees is small and their behaviour is sluggish, disease should be suspected and it may be necessary to get rid of the colony if this is found to be so.

Subsequent management

Depending upon the weather and the availability of food, the bees may be so active during the season that several supers will be needed. Regular examination will reveal when a new one is needed, and about once a week or every ten days is advisable during the late spring and early summer period. These examinations will also help to prevent swarming.

If new queen cells are formed, the likelihood of the colony swarming is greatly increased. Queen cells can be spotted fairly easily because they are larger than normal brood cells and stick out at an angle to the rest of the comb. Many bee-keepers destroy these with the hive tool when they come across them. It should be mentioned, however, that swarming can take place anyway, for it depends upon a number of factors, the weather, the availability of food and upon the inherent tendency of particular strains of bee to swarm. Specialist bee breeders are trying increasingly to breed strains of bee that have less of a tendency to swarm, and this should be borne in mind when purchasing new stock.

If a new colony is needed, an empty hive consisting of a brood chamber with some old brood comb can be used. Find a frame with queen cells in the existing hive and place it in the new hive. If there is more than one queen cell, only one should be left, and the others destroyed. There should be quite a lot of bees on the frame, but if there are not enough, some more can be shaken off other frames, and foundation brood frames should be placed in the brood chamber for them to work on.

The honey harvest

It is not a good idea to take the early honey because this will be depriving the colony when it is still expanding and building itself up in the early part of the summer. It is better to wait for the main honey flow, which is said to begin with the clover and to end with the lime blossoms. There is obviously a considerable seasonal variation and local conditions will need to be taken into consideration. In a good year, there will be so much honey that harvesting may take place several times.

All the equipment should be got ready first and, in addition to an extractor, strainer and settling tank, a sturdy working table, capping knife, trays and jars with screw-top lids will be needed. The room where the work is to be carried out should be adequately sealed off from the outside so that you are not bothered by stray bees or wasps.

The honey that the worker bees have produced in the super frames is there for winter feeding for the colony. A full National hive super will contain approximately 13.5 kg (30 lb), and if there are several supers in a good year it will be seen that the yield per hive can be high. Honey that is taken must be replaced by sugar solution or the colony will die in winter. Many bee-keepers leave a certain proportion of honey behind for the bees as well, on the basis that a colony which is honey-fed is stronger and healthier. There is also the feeling amongst many bee-keepers that to share the honey is fairer when one considers how hard the bees have worked to produce it and that it is ultimately better to work with nature than merely to exploit it.

Honeycombs which are ready for extraction are sealed, that is, each individual cell is capped. Honey which is taken from unsealed combs is not ripe and will ferment in the jars. The time to take the honey is when sealing has taken place and the honey flow has ceased. When this happens, place an escape board between the brood chamber and the first super. This is a contraption which allows the worker bees to go down to the brood box, but not to return to the super. In a day or two they will have gone down and the task of removing the supers is made easier. There will still be some bees, of course, but after a puff of smoke they can be shaken off into the hive.

As each frame is removed from the super, it should be replaced with an empty one so that the bees are encouraged to keep on working. Great care should be taken when moving the sealed honey frames, for a knock can damage the cappings, causing honey to leak out everywhere. The filled frames should be transferred to an empty super nearby so that the whole thing can then be carried indoors. Remember, however, that a full super can weigh between 13.5 and 18.1 kg (30–40 lb), so it is easier for two people to carry it. Once the supers have been removed and restocked with empty foundation frames, the escape board should be removed from the top of the brood chamber so that the bees have access to the supers again.

Before the frames are put in the extractor, the combs need to be uncapped. This is achieved by standing the frame on a tray and cutting upwards with a capping knife. The frame is then turned and the operation repeated on the other side. It is then placed in the extractor. Bee-keeping equipment suppliers sell electric capping knives which simplify the operation, but a long sharp knife dipped in hot water will perform the operation satisfactorily.

The extractor works by centrifugal force, with the honey being separated from the comb and collecting in a tank underneath. This has a tap so that the honey can be run off into a settling tank so that the air bubbles have a chance to rise to the surface before bottling takes place. Before settling, however, the honey must be strained to remove any wax, and the easiest way of doing this is to place a straining cloth of muslin or nylon over the top of the settling tank before the honey is allowed to run in. Leave to settle for a day or two, ensuring that the tub is covered to keep out dust and insects, then bottle in screw-top jars. If the honey is to be sold, the quantity must be stated on the label, and must be accurately measured, otherwise the weights and measures regulations will be contravened. It is also necessary to have the name of the supplier on the label. Most bee-keeping societies sell attractive labels to which the bee-keeper can add his own name.

Dealing with the wax

After the honey harvest, part of the clearing-up operation is collecting together all the bits of wax and recycling them. The cappings will still have some honey mixed with them and this can be reclaimed during the filtering process. There are several ways of dealing with the wax. The simplest is to put the cappings and bits and pieces in a large saucepan with about a pint of water and heat gently until the wax melts. On no account should it boil. Once all the wax has melted, put the pan aside to cool. The wax will form a cake and shrink away from the sides, making it easy to remove. It should then be gently washed and stored. The liquid is a valuable feed for any bees that are showing signs of food shortage.

Wax from other sources such as the scrapings from clearing operations on the hive, or old brood combs can be rendered by the use of solar power. An aluminium dish such as that used for cooking meat is placed at an angle of 45° in an insulated box. The latter can be a wooden box lined with old polystyrene tiles or pieces of polystyrene packaging. At the lower end of the dish is a container to catch the melted wax. Two sheets of glass, with 12.5 mm ($\frac{1}{2}$ in) space between them, are placed over the box and the whole thing then turned to face the sun.

Using the wax

Beeswax is a valuable commodity and has several uses, some of which can produce a profitable sideline.

Making foundation for new frames is an obvious use of recycled wax, for the cost of foundation frames is now high. Some bee-keepers

A home-made solar wax extractor

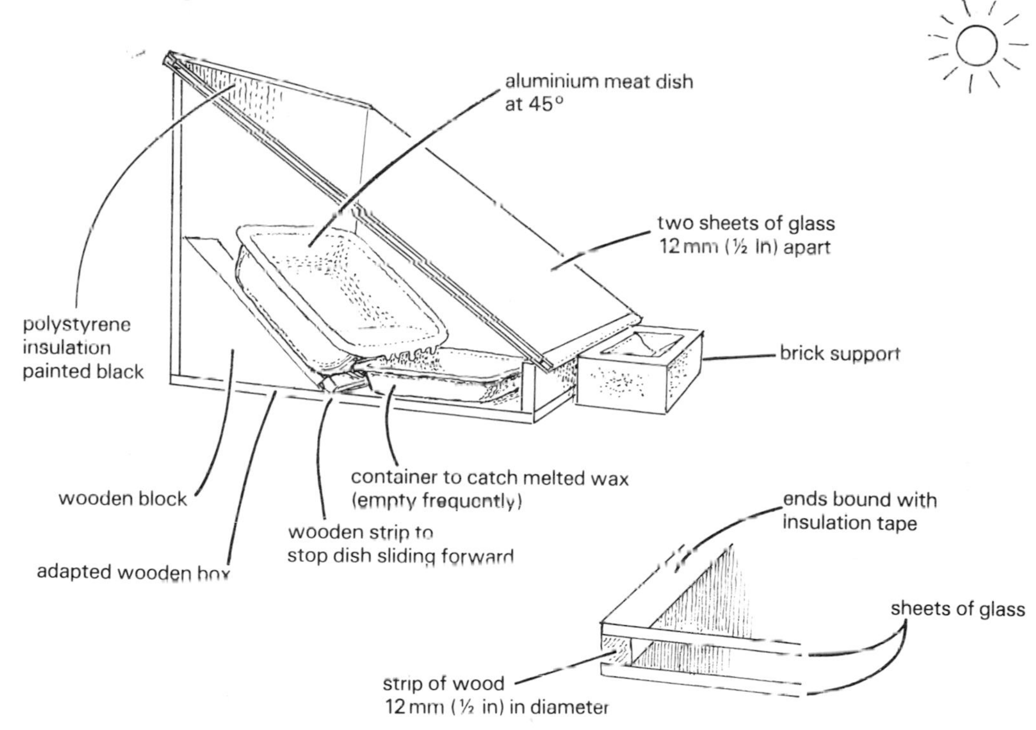

send their wax to appliance manufacturers to be made into foundation sheets and this is more costly than it is to make your own, but the finished product is of a high standard. It is possible to make your own mould by pouring plaster of Paris over a commercial sheet of foundation and then using that as a pattern for making future ones, but it does necessitate making a frame to hold the plaster while the initial mould is being prepared. Commercial moulds are also available.

The easiest way to make candles from beeswax is to buy plastic moulds and wicks from a crafts supplier. Place the mould upside down with a wick going through the hole at the bottom, and plug the hole with a small piece of putty. Tie the other end of the wick to a pencil resting across the open end of the mould. Place the beeswax in a waterbath to avoid over-heating, and heat until completely melted. Pour the wax into the moulds and stir slightly to ensure that there are no air bubbles, which could cause 'pitting' in the finished product. Ensure that the wick is in a central position, then leave to

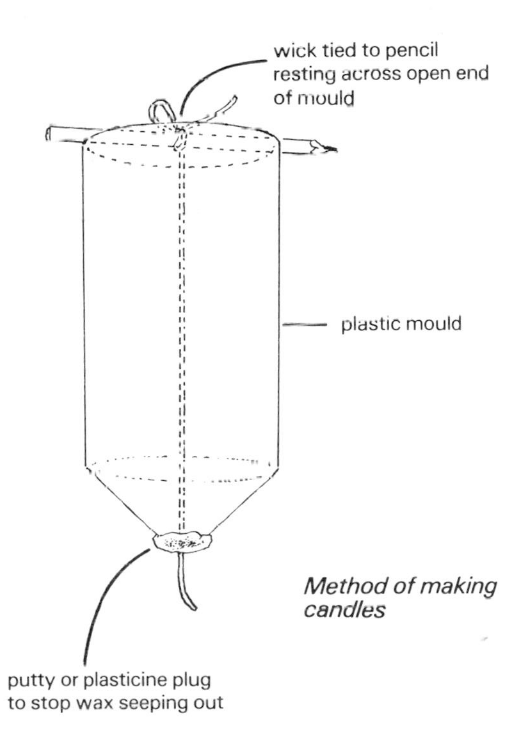

Method of making candles

cool. When quite cold the candles can be removed from the moulds. If this is difficult, dip them in water for a few minutes to loosen the wax. Candle moulds come in a variety of shapes and sizes, and most craft shops also sell purpose-made dyes should you wish to experiment with the production of coloured candles. Water baths can also be purchased or, alternatively, a dish placed in a saucepan of water will suffice.

Antique dealers and those with an interest in fine furniture are often interested to buy beeswax polish, which is preferable to the modern, silicone-based ones. It is easily made as long as appropriate containers for the finished product are obtained. It is also essential to use pure turpentine, rather than one of the substitutes sold in decorators' shops. Good D-I-Y suppliers will stock it, as well as those selling artists materials.

To make polish, melt beeswax in a waterbath until liquid, and remove from the heat. Extinguish all naked flames and quickly stir in pure turpentine. 0.5 litre (2 pt) will be needed for every 450 g (1 lb) of beeswax. When thoroughly mixed, pour into suitable containers and leave to cool. When cold, put lids on the containers and label them, 'Pure beeswax furniture polish'. The polish should not be difficult to sell.

Autumn management of hives

After the honey and wax harvests, it is important to check the hives and make preparations for the winter. There may have been several consecutive honey harvests depending upon how good the season was. Wet combs from which honey has been extracted can be returned to the hives for a few days, to be cleaned and repaired by the bees. Then they should be removed, wrapped carefully to preserve them from attack by the wax moth, and stored until needed next season.

Now is a good time to check the brood chamber and if possible to find the queen and check that she is healthy. She may have stopped laying at this time and if there is no indication of brood, it may be necessary to replace her. At this time of year, replacement queens are comparatively cheap and may be purchased from bee breeders. If the existing queen is healthy but not laying, she can often be induced to start by replacing the central brood frame with a foundation frame. Some of the brood frames may also have honeycombs in them, and this is something to be taken into account when autumn feeding is carried out.

From September onwards, concentrate on feeding until the colder weather starts. Make a sugar solution of 900 g (2 lb) white granulated sugar dissolved in 0.5 litre (1 pt) of hot water, allow to cool and place in the feeder. Replace as soon as it becomes low, to ensure that the bees are having enough before the cooler weather slows down their activity. Continue feeding until it becomes obvious that they have lost interest, then check that the hive itself is sturdy and free of roof leaks. Reduce the size of the entrance hole in order to keep out larger predators and, if necessary, insert a mouse guard. If high winds are likely to occur, place bricks on the roof in order to weigh it down. When snow falls, it is good practice to place a board in such a way as to cast shade on the entrance. This is to stop reflected light getting into the hive and possibly luring bees outside, where the cold would kill them. In the cold, winter months, the hives should not be opened otherwise the temperature fluctuations could have disastrous results.

17 HOME BUSINESSES TO ACCOMPANY PART-TIME FARMING

A small business that is home-based will have low overheads and will dovetail well with a part-time farming enterprise. There is an infinite variety of possibilities, depending upon individual interests and consumer needs, but it is worth considering, again, the effects that new technology will have on industry, and on employment trends. There will be increased leisure, with more emphasis being put on the D-I-Y, crafts and educational fields. Experts are also forecasting that the service industries will expand rapidly and that specialist shops and restaurants will increase. Repair industries are also expected to boom. There has been a major emphasis on the 'throwaway' aspect of consumer purchasing in recent years, with a parallel decline in repair services. All of us have experienced the frustration of not being able to find anyone locally who is either willing or capable of carrying out basic repairs to a range of products, from kettles to washing machines.

There is tremendous scope for making the best opportunities of what the future has to offer. Small home- and rural-based businesses will not only bring needed employment to those out of work, but will bring back life to rural communities. Many villages are, at present, little more than dormitories for town-working commuters, or retirement homes for old people. There is also the important fact that the more small businesses and independent enterprises there are in existence, the greater the hopes for democracy to survive.

The following is a list of avenues worth exploring as possible sources of revenue, in addition to the farming enterprises detailed in the book:

D-I-Y suppliers
Craft suppliers
Bicycle repairs
China repairs
Specialist bookshop
Window-cleaning agency
Garden maintenance agency
Woodburning supplies
Wholefood and real food shop
Livestock and pet feed suppliers
Sewing machine repairs
Radio and television repairs
Upholstery supplies and renovations
Typing or typesetting agency
Proof-reading service
Herbs and herbal products
Freelance writing
Freelance computer programming
Specialist cooking services
Photography services
Bookbinding
Picture framing
Bed and breakfast
Hiring agency
Book-keeping services
Boarding animals
Graphics service
Book indexing
Home dressmaking
Specialist courses and tuition
Mail order
Specialist livestock breeder
Small farming supplies shop
Specialist cooking supplies

Examples of part-time farming enterprises

The following are examples of part-time farming enterprises which I have seen in different parts of the country. There are many more and this is only a representative selection. It should be emphasized once again that a complete income from a relatively small piece of land is highly unlikely, and all the examples described rely on combinations of different income sources.

Mr J was made redundant several years ago and, as he was on the wrong side of forty-five, could not find another job. He was fortunate to find a run-down farmhouse with good barns and 6 acres of land. It was possible to buy this from the proceeds of the sale of his previous house in the town. He is a practical man with a wide range of D-I-Y- skills, and has been able to carry out most of the house repairs and alterations himself. His redundancy payment contributed most of the cost of these renovations.

His wife has professional qualifications and works in the nearby town for three days a week. Mr J works on the farm, full-time, and sells free-range eggs from a flock of 100 layers. These are sold 'at the farm gate', and also through local wholefood shops. Every Christmas he sells about 100 freshly killed turkeys to local buyers, who now supply regular seasonal orders. He buys the turkey poults in August and houses them in a large barn on straw. Keeping calves on a contract basis is another of Mr J's income sources. He is paid so much for each calf he raises to a certain weight. He also has a small pig unit where weaners are raised to pork weight for sale. He keeps one or two to bacon weight for his own use and also keeps a house cow for the family's dairy needs. He has acquired a lot of second-hand machinery and does his own ditching, hedging and hay-making. Mr J grows a certain proportion of his own fodder crops and hay, but also buys proprietary feedstuffs in bulk. With plenty of suitable barns, he is able to store these without difficulty, and has now started a small business, selling livestock feeds in smaller quantities to local smallholders and goat-keepers.

Mr and Mrs B used to live in a tied cottage on a large farm estate. Mr B worked three days a week in the intensive pig unit, and with the sheep which represented the second of the farm's enterprises. They were allowed to rent grazing land from the farm and built up a herd of twenty milking goats. They sold goat's milk and yoghurt at the gate and through local shops but as demand grew, so did the size of the herd, and soon they were supplying a number of retail outlets further afield. They have now bought their own farm of 10 acres and have increased their herd to over a hundred goats. They have invested in a pipeline milking unit, as well as a commercial yoghurt vat, and Mrs B makes yoghurt to supply over fifty retail outlets. Mr B spends two and a half days delivering the produce to the shops, and although he had a part-time job working for another farmer, he has now given this up to concentrate on his own enterprise. They employ no outside help although they receive some help from their teenage children. They say that they are happy with what they are doing, but that it is 'virtually a twenty-four-hour job'.

Mr W has a 50-acre farm, which used to be a dairy enterprise and has good grazing land. He concentrates on bullocks, for the prime beef market, rearing them until they are about twenty months old. He has planted 5 acres of his land with trees as a commercial timber crop, and keeps chickens and a few sheep for his own family use. His main income source comes from his small scientific consultancy business, which operates from a converted outbuilding on his land. He employs a part-time secretary for the business and casual labour, when it is necessary, for managing the bullocks. He utilizes contract labour for hedging, ditching and hay-making.

Mr P is a builder with a share in a business. He and his wife have 4 acres of land and keep Jacob sheep, as well as chickens for their own needs. They have adapted one of their outbuildings as a rearing house for table ducks, and buy in 100 day-old ducklings at a time. These are reared to table weight and then sold through the poultry trade.

Mrs P developed a skill at curing her Jacob sheep skins, and has now started a small home business, curing skins for local sheep and goat owners.

Mrs C and her husband have a farmhouse and 6 acres. Mr C works in the City of London and his wife looks after three Jersey cows, some sheep, pigs and poultry. One of the barns has been converted to a wholefood shop and a regular trade has been built up for wholefoods and free-range eggs. In addition, Mrs C has applied for, and been granted, a cream licence so that she is able to sell her own Jersey cream. Recently, she formed a partnership with Mr S, a freelance butcher, to sell meat, and they now have their own licensed butcher's shop next to the wholefood shop. They specialize in 'real' meat untainted by antibiotics or steroids and sell their own beef, lamb and pork as well as that bought locally 'on the hoof', from other organic farmers in the area. The animals are slaughtered and inspected at the local abattoir and then returned to their butcher's shop for cutting up. They employ part-time, casual labour to help with the day-to-day farming activities.

Mrs B is a qualified typesetter and does typesetting at home on a freelance basis. Her husband is an accountant, working full-time for a local firm. They have 3 acres of land and grow most of their own fruit and vegetables. They keep chickens and ducks, mainly for their own use, but sell surplus free-range eggs at the gate. They buy in lambs every spring and graze them through to the autumn, when they are sold to a local butcher. They keep two for their own freezer. They have a large fruit garden which produces top quality raspberries. The crop is sold to a wholesaler every year, leaving just enough for their own use. They employ a part-time gardener for a few hours a week.

Mr and Mrs M have a house with one acre of land on which they keep pure-bred ducks. They sell duck eggs through a local shop and also breed ducklings for sale. They grow their own vegetables, and raise two lambs a year for the freezer. Surplus vegetables are sold at the gate.

Their main source of income comes from their home-based business selling woodburning stoves. They have converted a barn into a showroom, and act as local agents for a number of woodstove manufacturers. They also have a number of mechanical log-splitting devices which they hire out to customers. They do not employ anyone, and do all the work on the business and on the land themselves.

Mr P is a bee-keeper of many years standing. He first started it as a hobby while he was working full-time in the catering trade. Now that he has retired, he has put his bee-keeping activities on a more commercial footing. He not only sells his honey to callers and through wholefood shops, but has become a local agent for a firm of bee-keeping equipment manufacturers. He has just under 3 acres of land and has converted an outbuilding near his house, into a small office and shop where bee equipment is on view. In addition to this, he has a small plantation of Christmas trees, and each year local customers come and select their own tree. His wife makes a selection of soft toys and craft items for the Christmas season, and these are on view in the shop at this time. They keep chickens and grow vegetables for their own use, but occasionally offer surplus produce for sale.

Mr H has a 45-acre dairy farm in a scenic part of the country. He sells liquid milk from his herd of fifty cows through the Milk Marketing Board, and this provides his basic income. As his farm is in a tourist area, he has turned part of it into a farm park. There is a signposted walk with the various features described for the benefit of visitors. He also keeps a variety of rare breeds of livestock and poultry, purely as a tourist attraction. At the end of the walk is a small café providing snacks and a range of gift items. This is run and manned by his mother. He employs one man to help him on the farm, and utilizes student labour to help with the farm park in the tourist season.

Mr and Mrs T have a 30-acre hill farm, with a commercial flock of sheep. The lambs are sold to the meat trade, while fleeces are taken by the Wool Marketing Board. A certain proportion of fleeces are used for spinning demonstrations at a small exhibition centre in one of the converted barns on their farm. Here they run courses in spinning, weaving and natural dyeing, as well as selling appropriate equipment and books on the subject. Mrs T is also employed as a part-time school teacher, teaching crafts, as well as giving night school classes for the local authority in the winter months. Their own courses and demonstrations take place mainly in the summer holidays and are a popular tourist attraction. Mr and Mrs T also keep chickens and grow vegetables for their own use. They use part-time student help in the summer months only, and use contract labour for hedging, ditching and shearing.

Mr T is a policeman with $\frac{1}{2}$-acre garden and an attached paddock of about 1 acre. The garden is mainly given over to vegetables, which

Mr T tends in his off-duty hours. He quite enjoys doing night duty because he says that it means more daylight hours working in the garden, but he did not specify how much sleep he got.

Surplus vegetables are sold at the gate by his wife. Free-range eggs are also on offer from a small flock of chickens kept in a run at the bottom of the garden. In winter they have a shed equipped with electricity to keep up the egg yield. The field is used as pasture for two dairy goats and the occasional lamb reared for the freezer. Goat's milk and yoghurt are sold at the gate, as well as the occasional soft cheese.

Mr T grows a certain proportion of winter fodder crops for the goats. This is mainly in the form of thousand-headed kale which is sown in June and planted out in July. The advantage of this variety is that you can keep on cutting it and it will produce more heads.

Mss O and T have 10 acres of land in a hilly and scenic area of the border country. There is an old manor house with nine bedrooms, which has been converted to cater for parties of twelve children plus staff. The Centre accepts groups of children from schools or other agencies, and also runs an established scheme in co-operation with the National Association of Gifted Children for holidays for children of outstanding ability. City children who may be disadvantaged materially are also catered for. There are three full-time resident staff augmented at busy times with teachers, students and outside helpers on the land and with the animals.

There are various projects for the children to take part in, such as helping with the daily running of the smallholding; learning to make cream, cheese and bread; making shelters and campfires and cooking their suppers on them; and studies of local castles and wildlife.

The Centre has a range of livestock, including poultry, Tamworth pigs, Dexter cattle and British Toggenburg goats. The cows and goats provide milk for the farm's dairying needs and the pigs provide weaners and pork pigs, which are sold through the meat trade. A kitchen garden provides a proportion of the Centre's fresh vegetables.

Mr and Mrs S have an 8-acre farm, which is run on a part-time basis with the main emphasis being on store cattle. In addition, Mr S has a part-time job as a milk recorder for the Milk Marketing Board. This involves visiting local dairy farms once a month for an evening and following morning milking, and taking milk samples from each cow so that the yield and butter-fat content can be recorded. The service is provided by the MMB and the farmers pay, while Mr S operates on a contract basis. Mr S says that a reliable car is needed, one that won't break down at 5 am when he is setting out for a morning milking. It is a job for the early riser and the late caller, so it may not suit many people, but the advantage is that the main part of the day is clear for concentrating on his own farm.

He keeps free-range hens, as well as a small number of sheep that are normally bought in as lambs each year and then kept until large enough to sell through the meat trade. A kitchen garden provides much of Mr and Mrs S's fruit and vegetable needs, and they sell surplus eggs and tomatoes at the gate. Mrs S says she has often considered having a dairy animal herself but thinks that it would be too much of a tie, and that her husband would not relish having to milk it, because it would be too much like his work. Mr S is able to buy milk direct from the farm whenever his wife wants to make cheese, so there is really no need for their own dairy animals.

Mr and Mrs G have a beautiful seventeenth-century house with 6 acres of land. There is a walled kitchen garden, an orchard and several paddocks. They keep a flock of laying hens and some Jacob sheep.

Mr G works full-time as the director of a charitable foundation. Mrs G runs a wholefood and traditional food shop in a converted building adjacent to the house, where she sells a lot of their own produce, including free-range eggs, herbs, fruit and vegetables, and traditional marmalades, pickles and other specialities that Mrs G makes in her own kitchen. The Jacobs are kept as a breeding flock, with the lambs either being sold to local buyers or slaughtered for the freezer. The fleeces are used by Mrs G, who is actively interested in spinning.

Mr G works for the council and lives in a house with a small garden. He has no land of his own, yet still manages to deal in sheep. He rents odd pieces of grazing land from a number of farmers and buys young lambs in the market. These are grazed to meat weight and then sold to butchers for slaughtering at a local abattoir. He has a trailer with tailboard which he tows behind his car, and this is used for transporting the stock. Mr G says that his rental costs are low and he does not have any overheads, but his petrol costs are high. Even so, he manages to make a reasonable profit on the sale of his meat lambs.

Mr and Mrs R have a house and 6 acres in the Thames basin area, so that they are near to a number of large towns. They are fond of livestock and are keen crafts people. All these factors have led them to turn their small farm into an exhibition centre for parties of schoolchildren. There is a wide variety of livestock, including a Jersey cow, goats, sheep, pigs, pony, donkey and poultry.

The exhibition is organized in the form of a 'trail' that children follow, equipped with project papers which they can consult as they proceed. The trail ends in a building where craft demonstrations and exhibits of spinning, weaving and dyeing are shown. There is also a small shop selling the farm's own poster and postcards, and there is a protected picnic area. They have an average of 3,000 children a year and the enterprise is largely run by Mrs R, while her husband has a full-time job as a civil servant.

Mr G is a retired arable and stock farmer who did not want to relinquish his connection with stock entirely, and yet did not wish to have an occupation which took up more than two or three hours a day. He and his wife run a 100-doe rabbit unit, which provides an income from the sale of meat rabbits to the rabbit packers. The rabbits are sold live and are collected by the packers.

Mr K lives in a tied cottage working as a chauffeur and general handyman on a large estate. He has the use of 1 acre of land and breeds ornamental waterfowl, concentrating on the more expensive types. He has invested in a small incubator to ensure maximum hatchings, and reckons that he does better out of his ducks than from his employment activities.

FURTHER INFORMATION

Publications

MAGAZINES

Practical Self Sufficiency, Broad Leys Publishing Company, Widdington, Saffron Walden, Essex CB11 3SP. (UK)

Countryside and Small Stock Journal, Rt. 1, Box 239, Waterloo, Wisconsin 53594. (USA)

Mother Earth News, Box 70, Hendersonville, NC 28739. (USA)

Grass Roots, Box 900, Shepparton 3630. (Australia)

Earth Garden, PO Box 378, Epping, New South Wales 2121. (Australia)

The Small Farmer, PO Box 2081, Palmerston North. (New Zealand)

Harrowsmith, Camden House Publishing Ltd., Camden East, Ontario KOK 1JO. (Canada)

Le Pont, Editions de la Laterne, 5 Rue du Lac, Magny Vernois, 70 200 Lure. (France)

BOOKS

The Complete Book of Raising Livestock and Poultry, Katie Thear and Dr Alisdair Fraser (London, 1980)

The Complete Book of Self Sufficiency, John Seymour (London, 1976)

Backyard Farming, Ann Williams (London, 1978)

Making and Managing a Smallholding, Michael Allaby (London)

The Smallholder's Guide, C. J. Munroe (London, 1979)

Successful Small Farming, Malcolm Blackie (New Zealand, 1981)

Living on a Little Land, Patrick Rivers (London, 1978)

The Smallholder's Year, Katie Thear (London 1981)

Farm your Garden, Joanna Smith (London, 1977)

Cottage Economy, William Cobbett (London, 1979)

The Old Fashioned Recipe Book: An Encyclopedia of Country Living, Carla Emery (1977)

Raising your own Livestock, Claudia Weisburd (USA, 1980)

The Home Dairying Book, Katie Thear (London, 1978)

Practical Milk Production, David Morris (London, 1976)

Keeping Livestock Healthy, N. Bruce Haynes (USA)

TV Vet Book for Stock Farmers, (Books 1 & 2) (London)

Profitable Beef Production, M. M. G. Cooper & M. B. Willis (London, 1972)

Goat Husbandry, David Mackenzie (London, 1957)

A Practical Guide to Small Scale Goat Keeping, Billie Luisi (USA, 1979)

Practical Rabbit Keeping, Katie Thear (London, 1981)

Raising Rabbits, Ann Kanable (USA, 1977)

The Complete Handbook of Poultry Keeping, Stuart Banks (London, 1979)

Raising Poultry the Modern Way, Leonard S. Mercia (USA, 1975)

Raising the Home Duck Flock, David Holderread (USA, 1978)

Bantams for Everyone, H. Easom Smith (London, 1976)

Keeping Domestic Geese, Barbara Soames (London, 1980)

Raising Your own Turkeys, Leonard S. Mercia (USA, 1981)

Guinea Fowl, Van Hoesen (USA, 1975)

Sheep Management and Production, Derek Goodwin (London, 1979)

Raising Sheep the Modern Way, Paula Simmons (USA)

The Book of the Pig, Susan Hulme (London, 1979)

Small-Scale Pig Raising, Dirk van Loon (USA, 1978)

Organic Gardening under Glass, George and Katy Abraham (USA, 1976)

The Vegetable Garden Displayed, The Royal Horticultural Society (London, 1975)

The Fruit Garden Displayed, The Royal Horticultural Society (London)

Horticulture in Britain, H.M.S.O. (London, 1971)

The New English Vineyard, Joanna Smith (London, 1979)

A Modern Herbal, Grieve and Leyel (London, 1974)

The Rodale Herb Book, (USA, 1976)

The Illustrated Earth Garden Herbal, Keith Vincent-Smith (Australia, 1979)

Growing and Using Herbs Successfully, Betty Jacobs (USA, 1981)

Australia & New Zealand Organic Gardening, Peter Bennet (Australia, 1979)

The Complete Handbook of Bee-Keeping, Herbert Mace (London)

A Guide to Keeping Bees in Australia, Norman Redpath (Australia, 1981)

Practical Beekeeping, Tompkins & Griffith (USA, 1977)

Fertility Gardening, Lawrence Hills (London, 1980)

Tree Nurseries, K. Liebscher (London, 1979)

Earning Money at Home, Consumers Association (London, 1979)

Working for Yourself, Geof. Hewit (USA, 1977)

The Ministry of Agriculture, Fisheries and Food produces a booklet entitled *At the Farmers Service*, which lists the services available to farmers. They also have a list of free publications which will be sent on application. The department of agriculture of other countries offer a similar service and the addresses are listed in the appropriate sections.

United Kingdom

ORGANIZATIONS

Ministry of Agriculture, Fisheries and Food (MAFF), Whitehall Place, London SW1A 2HH.
Agricultural Development and Advisory Service (ADAS), Great Westminster House, Horseferry Road, London SW1P 2AL. (For divisional offices, see local *Yellow Pages*.)
Department of Agriculture and Fisheries for Scotland, Chesser House, Gorgie Road, Edinburgh EH11 3AW.
Organic Farmers and Growers Ltd, 9 Station Approach, Needham Market, Ipswich, Suffolk IP6 8AT.
Organic Growers Association, Aeron Park, Llangeitho, Tregaron, Dyfed, Wales.
The Soil Association, Walnut Tree Manor, Haughley, Stowmarket, Suffolk IP14 6RS.
The Henry Doubleday Research Association, 20 Convent Lane, Bocking, Braintree, Essex.
National Vegetable Research Station, Wellesbourne, Warwick.
Association of British Herb Growers and Producers, The Herb Centre, Middleton Tyas, Nr Richmond, Yorkshire.
British Agricultural and Garden Machinery Association Ltd, 14–16 Church Street, Rickmansworth, Herts WD3 1RQ.
British Trust for Conservation Volunteers, 10–14 Duke Street, Reading, Berks.
Horticultural Trades Association, 18 Westcote Road, Reading, Berks.
Scottish Seed and Nursery Trade Association, 14 Bruntsfield Crescent, Edinburgh EH10 4HA.
Council for the Protection of Rural England, 4 Hobart Place, London SW1W 0HY.
English Vineyards Association Ltd, The Ridge, Lamberhurst Down, Kent TN3 8ER.
National Vegetable Society, 29 Revidge Road, Blackburn, Lancs BB2 6JB.
Pot Plant Growers Association, Tanner Farm, Goudhurst Road, Marden, Kent TN12 9ND.
Tree Council, 35 Belgrave Square, London SW1X 8QN.
Men of the Trees, Crawley Down, Crawley, Sussex.
The Forestry Commission, Information Branch, 213 Corstorphine Road, Edinburgh EH12 7AT.
Milk Marketing Board, Thames Ditton, W. Surrey.
National Dairy Council, 5 St John, Princes Street, London W1.
The Cheese Bureau, 40 Berkeley Square, London W1X 6AD.
Butter Information Council, Bank Street Suite, 158 High Street, Tonbridge, Kent TN9 1BJ.
National Federation of City Farms, 15 Wilkin Street, London NW5.
Working Weekends on Organic Farms (WWOOF), 19 Bradford Road, Lewes, Sussex BN7 1RB.
The Jersey Cattle Society of Great Britain, 154 Castle Hill, Reading, Berks.
British Friesian Cattle Society of Great Britain & Ireland, Scotsbridge House, Rickmansworth, Herts WD3 3BB.
The British Goat Society, Rougham, Bury St Edmunds, Suffolk IP30 9LJ.
Rare Breeds Survival Trust, 4th Street, National Agricultural Centre, Stoneleigh, Kenilworth, Warks CV8 2 LG.
The National Pig Breeders Association, 49 Clarendon Road, Watford, Herts.
National Sheep Association, Jenkins Lane, St Leonards, Tring, Herts.
Meat and Livestock Commission, Queensway House, Bletchley, Milton Keynes, Bucks.
British Wool Marketing Board, Oak Mills, Clayton, Bradford, W. Yorks.
Grassland Research Institute, Hurley, Maidenhead, Berks.
Hill Farming Research Organization, Bush Estate, Penicuik, Midlothian.
National Institute of Agricultural Botany, Huntingdon Road, Cambridge.
The British Poultry Federation, 52–54 High Holborn, London WC1V 6SX.
British Egg Information Service, 37 Panton Street, London W1.
Free-Range Egg Society (FREGG), 39 Maresfield Gardens, London NW3.
The British Turkey Federation, High Holborn House, 52–4 High Holborn, London WC1V 6SE.
British Waterfowl Association, Market Place, Haltwhistle, Northumberland NE49 0BL.
Duck Producers Association, High Holborn House, London WC1V 6SE.
British Rabbit Council, Purferoy House, 7 Kirkgate, Newark, Notts.
Commercial Rabbit Association, Tyning House, Shurdington, Cheltenham, Glos.
British Beekeepers Association, 55 Chipstead Lane, Sevenoaks, Kent TV13 2AJ.
Scottish Beekeepers Association, 26 The Meadows, Berwick-on-Tweed, Northumberland TD15 1NY.
Bee Research Association, Hill House, Chalfont St Peter, Gerrards Cross, Bucks SL9 0NR.
British Isles Bee Breeders Association, Whitegates, Thulston, Derby DE7 3EW.
Registrar of Business Names (England and Wales), Pembroke House, 40–50 City Road, London EC1; (Scotland) George Street, Edinburgh LH2 3DJ; (Northern Ireland) 43–7 Chichester Street, Belfast BT1 4RJ.

Small Firms Information Centre (London Region), 65 Buckingham Palace Road, London SW1W 0QX. Other regional centres are listed in local telephone directories.

National Federation of Self-Employed and Small Businesses, 32 St Anne's Road West, Lytham St Annes, Lancs FY8 1NY.

Council for Small Industries in Rural Areas (CoSIRA), Queen's House, Fish Row, Salisbury, Wilts SP1 1EX.

Federation of British Craft Societies, 43 Earlham Street, London WC2H 9LD.

Arts Council of Great Britain, 105 Piccadilly, London W1V 0AU.

Guild of Weavers, Spinners and Dyers, Five Bays, 10 Stancliffe Avenue, Morford, Wrexham, Clwyd.

Woodburning Association of Retailers and Manufacturers, PO Box 35, Stoke-on-Trent, Staffs ST4 7NU.

British Tourist Association, 64 St James's Street, London SW1.

The Caravan Club, Grinstead House, London Road, East Grinstead, W. Sussex.

National Trust, 42 Queen Anne's Gate, London SW1.

Ecology Building Society, 43 Main Street, Crosshills, via Keighley, W. Yorks BD20 8TT.

SUPPLIERS

Self Sufficiency and Smallholding Supplies, Wells, Somerset BA5 1SY.

Earthworks, 22 Corve Street, Ludlow, Shropshire.

Delrose (Mail Order Supplies), The Gate House, Millstone Lane, Oakerthorpe, Derbyshire DE5 7LP.

Lakeland Plastics, 52 Alexandra Buildings, Station Precinct, Windermere, Cumbria.

Transatlantic Plastics Ltd, Ventnor, Isle of Wight.

Southern Pullet Rearers Poultry and Self Sufficiency Centre, Greenfields Farm, Fontwell Avenue, Eastergate, Chichester, W. Sussex.

ISA Poultry Services Ltd, Orton Longeville, Peterborough PE2 0DN.

East Anglian Rabbit (Marketing) Ltd, Heath Road, Hickling, Norwich NR12 0AX.

Hylyne Rabbits Ltd, Marston, Northwich, Cheshire.

Country Craft, 22 Market Street, Alton, Hants.

Gregory, Prentis and Green Ltd (Spinning and Weaving), Mace Lane, Ashford, Kent TN24 8PE.

E. H. Taylor Ltd, Beehive Works, Welwyn, Herts AL6 0AZ.

Robert Lee (Bee Supplies) Ltd, Beehive Works, George Street, Uxbridge, Middlesex UB8 1SX.

Down to Earth Smallholders Seeds, Streetfield Farm, Cade Street, Heathfield, E. Sussex.

J. W. Moles & Son, Seed Merchants, Stanway, Colchester, Essex.

Thompson & Morgan (Seeds) Ltd, London Road, Ipswich, Suffolk.

Rossendale Electronic Fencers, 25 Lumb, Rossendale, Lancs.

Wolseley Webb Ltd (Rotovators), Electric Avenue, Witton, Birmingham B6 7JA.

United States of America

ORGANIZATIONS

United States Department of Agriculture (USDA), Washington DC 20505.

Natural Organic Farmers Association of Vermont, Rd. 1, Box 30, Hardwick, Vermont 05843.

Maine Organic Farmers & Gardeners Association, PO Box 187, Hallowell, Maine 04347.

The American Dairy Goat Society, Box 186, Spindale, North Carolina 28160.

The American Jersey Cattle Club, 2105 J. South Hamilton Road, Columbus, Ohio 43227.

The Holstein-Friesian Association of America, Box 808, Brattleboro, Vermont 05301.

American Poultry Association, Box 70, Cushing, Oklahoma 74023.

American Bantam Association, PO Box 610, N. Amherst, Maine 01059.

International Waterfowl Breeders Association, 12402 Curtis Road, Grass Lake, Michigan 49240.

American Rabbit Breeders Association, 1007 Morrissey Drive, Bloomington, Illinois 61701.

American Bee Society, Hamilton, Illionois 62341.

SUPPLIERS

Mother's General Store, Box 506, Flat Rock, North Carolina 28731.

Countryside General Store, 312 Portland Road, Waterloo, Wisconsin 53594.

Sheepman's Supply Co., Rte 1, Box 141, Barboursville, Virginia 22923.

New England Cheesemaking Supply Co., Box 85, Ashfield, Maine.

Chr. Hansens Laboratory, Milwaukee, Wisconsin, 53214.

Johnny's Selected Seeds, Albion, Maine, 04910.

W. Atlee Burpee Co., 300 Park Avenue, Warminster, Pennsylvania 18974.

Canada

ORGANIZATIONS

Ministry of Agriculture Canada, 930 Carling Avenue, Ottawa, Ontario.

Ontario Dairy Goat Society, RR3, Crysler, Ontario.

The Canadian Cattle Breeders Society, Roxton Pond, Shefford, Quebec J0E 1Z0.

Canadian Jersey Cattle Club, 343 Waterloo Avenue, Guelph, Ontario N1M 3K1.

Holstein-Friesian Association of Canada, Brantford, Ontario.
Horticultural Research Institute, Vineland Station, Ontario.
Canadian Organic Growers, 33 Karnwood Drive, Scarborough, Ontario M1L 2Z4.

SUPPLIERS

Homestead Equipment, Box 339, Acton, Ontario L7J 2M4.
The Pioneer Place, Box 100, Ronte 4, Aylmer, Ontario N5H 2R3.
Canadian Poultry Supplies, Rt 2, Lindsay, Ontario K9V 4R2.
Hovan-Lally Co. Ltd, 1146 Aerowood Drive, Mississauga, Ontario L4W 1Y5.
Stokes Seeds Ltd, 39 James Street, Box 10, St Catharine's, Ontario L2R 6R6.
Frey's Hatchery, 70 Northside Drive, St Jacobs, Ontario, N0B 2N0.
Canadian Alternatives, Lomans & Stephen, RR1, Caledon East, Ontario L0N 1E0.

Australia

ORGANIZATIONS

Department of Agriculture and Fisheries (DAF), 25 Grenfell Street, Adelaide 5000.
The Goat Breeders Society of Australia, Box 4317, GPO Sydney, NSW 2001.
Forestry Commission of NSW, GPO Box 2667, Sydney.
The Society for Growing Australian Plants, 860 Henry Lawson Drive, Picnic Point, NSW 2213.
Henry Doubleday Research Association of Australia, Sahara Farm, Robson Road, Kenthurst 2154.
Soil Association of South Australia Inc., GPO Box 2497, Adelaide 5001.
Organic Gardening and Farming Society of Tasmania Inc., GPO Box 228, Ulverstone 715.
Working Weekends on Organic Farms (WWOOF), 7 Duncan Avenue, Boronia, Victoria 3155.
Commercial Apiarists' Association, 88 South Street, Granville, NSW.
Victorian Apiarists' Association, PO Box 137, Noble Park, Victoria 3174.
The Tasmanian Farmers Federation, 54–6 Mount Street, Burnie, Tasmania 7320.
The Royal Agricultural and Horticultural Society of South Australia, Showground, Wayville, SA 5034.

SUPPLIERS

Self Sufficiency Supplies, 256 Darby Street, Newcastle 2300.
Alternatives, 37 Bangella Street, Torwood, Brisbane.
Going Solar, Energy, Agriculture and Self Sufficiency Supplies, 438 Queen Street, Melbourne 3000.
The Shearin' Shed Spinning and Weaving Supplies, 483 Crown Street, Surry Hills 2010, NSW.
Victorian Rennet Mfg Co. Pty, 16 Queen Street, Munawaging, Victoria 3131.
Henderson Seed Co., Box 118, Bulleen 3105.
Digger Seeds, 119 Ashworth Street, Albert Park 3106.
Tasmanian Forest Seeds, Summerleas Farm, Kingston, Tasmania 7150.
Eerindale Hatchery, 158 The Entrance Road, Erina, via Gosford 2250, NSW.
Beringa Stud Farm and Hatchery, 179 New Line Road, West Pennant Hills, 2120 NSW.

New Zealand

ORGANIZATIONS

Ministry of Agriculture, Private Bag, Wellington, New Zealand.
New Zealand Association of Small Farmers, PO Box 2081, Palmerston North.
Doubleday Research Association of New Zealand (DRANZ), PO Box 8843, Symonds Street, Auckland.
Craft Council Resource Centre, 110–116 Courtenay Place, Wellington.
Tree Crops Association, Crop Research Division, Lincoln, Christchurch.
Commission for the Environment, PO Box 10241, Wellington.
Community Enterprise Loan Trust, 8 Ngapuhi Road, Auckland 5.
National Beekeepers Association of New Zealand, Box 4048, Wellington.

SUPPLIERS

Mount Industries Ltd (Farm Buildings), 33 Mahana Road, PO Box 10039, Hamilton.
D. B. McIntosh (Farm Machinery), Willowby, No. 4 RD, Ashburton.

INDEX

Page numbers in italics refer to illustrations